An introduction to **nuclear physics**

W.N.COTTINGHAM
University of Bristol
D.A.GREENWOOD
University of Bristol

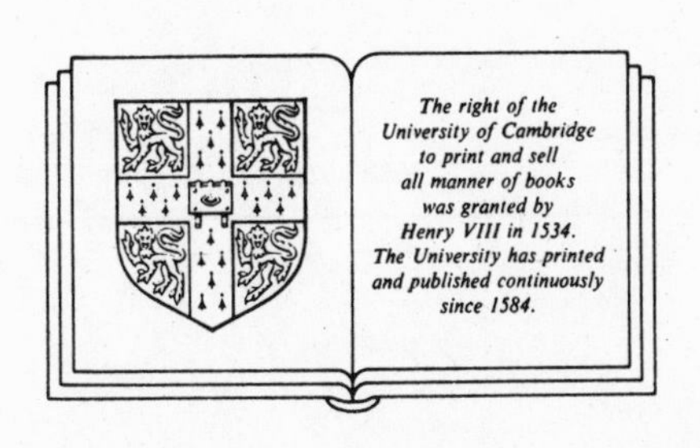

CAMBRIDGE UNIVERSITY PRESS
Cambridge
New York Port Chester
Melbourne Sydney

Published by the Press Syndicate of the University of Cambridge
The Pitt Building, Trumpington Street, Cambridge CB2 1RP
40 West 20th Street, New York, NY 10011, USA
10 Stamford Road, Oakleigh, Melbourne 3166, Australia

First published 1986
Reprinted 1987 (with corrections and additions), 1990

Printed in Great Britain at the University Press, Cambridge

Library of Congress cataloguing in publication data

Cottingham, W.N.
An introduction to nuclear physics.

Bibliography
Includes index.
1. Nuclear physics. 2. Particles (Nuclear physics).
I. Greenwood, D.A. II. Title.
QC776.C63 1986 539.7 85-22414

British Library cataloguing in publication data

Cottingham, W.N.
An introduction to nuclear physics.
1. Nuclear physics
I. Title II. Greenwood, D.A.
539.7 QC776

ISBN 0 521 26580 0 hardback
ISBN 0 521 31960 9 paperback

MP

Contents

Preface

In writing this text we were concerned to assert the continuing importance of nuclear physics in an undergraduate physics course. We set the subject in the context of current notions of particle physics. Our treatment of these ideas, in Chapters 1 to 3, is descriptive, but it provides a unifying foundation for the rest of the book. Chapter 12, on β-decay, returns to the basic theory. It also seems to us important that a core course should include some account of the applications of nuclear physics in controlled fission and fusion, and should exemplify the role of nuclear physics in astrophysics. Three chapters are devoted to these subjects.

Experimental techniques are not described in detail. It is impossible in a short text to do justice to the ingenuity of the experimental scientist, from the early discoveries in radio-activity to the sophisticated experiments of today. However, experimental data are stressed throughout: we hope that the interdependence of advances in experiment and theory is apparent to the reader.

We have by and large restricted the discussion of processes involving nuclear excitation and nuclear reactions to energies less than about 10 MeV. Even with this restriction there is such a richness and diversity of phenomena that it can be difficult for a beginner to grasp the underlying principles. We have therefore placed great emphasis on a few simple theoretical models that provide a successful description and understanding of the properties of nuclei at low energies. The way in which simple models can elucidate the properties of a complex system is one of the surprises of the subject, and part of its general educational value.

We have tried to keep the mathematics as simple as possible. We assume a knowledge of the basic formulae of special relativity, and some basic quantum mechanics: wave equations, energy levels and the quantisation of angular momentum. A few topics which may not be covered in elementary courses in quantum mechanics are treated in appendices. We consider the technicalities of angular momentum algebra, phase shift analysis and isotopic spin to be inappropriate to a first course in nuclear physics. Equations are written to be valid in SI units; results are usually expressed in MeV and fm. Each chapter ends with a set of problems intended to amplify and extend the text; some refer to further applications of nuclear physics. We have covered the bulk of the material in this book in 35 lectures of the core undergraduate curriculum at Bristol; these are given in the second and third years of the honours physics course.

We thank colleagues and students who read drafts of the text and drew our attention to errors and obscurities, which we have tried to eliminate. We are grateful to Margaret James and Mrs Lilian Murphy for their work on the typescript.

There is a less obvious debt: to the sometime Department of Mathematical Physics of the University of Birmingham where, under Professor Peierls, we first learned about physics.

W. N. Cottingham
D. A. Greenwood

Bristol, August 1985

Constants of nature and conversion factors

Velocity of light	c	2.99792×10^8 m s^{-1}
Planck's constant	$\hbar = h/2\pi$	1.05459×10^{-34} J s
Proton charge	e	1.60219×10^{-19} C
Boltzmann's constant	k_B	1.3807×10^{-23} J K^{-1} $= 8.617 \times 10^{-5}$ eV K^{-1}
Gravitational constant	G	6.67×10^{-11} m^3 kg^{-1} s^{-2}
Fermi coupling constant	G_F	$1.136 \times 10^{-11} (\hbar c)^3$ MeV^{-2}
Electron mass	m_e	9.1095×10^{-31} kg $= 0.51100$ MeV$/c^2$
Proton mass	m_p	1.007276 amu $= 938.28$ MeV$/c^2$
Neutron mass	m_n	1.00866 amu $= 939.57$ MeV$/c^2$
Atomic mass unit	(mass ^{12}C atom)/12	1.66057×10^{-27} kg $= 931.50$ MeV$/c^2$
Bohr magneton	$\mu_B = e\hbar/2m_e$	5.78838×10^{-5} eV T^{-1}
Nuclear magneton	$\mu_N = e\hbar/2m_p$	3.15245×10^{-8} eV T^{-1}
Bohr radius	$a_0 = 4\pi\varepsilon_0\hbar^2/m_e e^2$	0.529177×10^{-10} m
Fine-structure constant	$e^2/4\pi\varepsilon_0\hbar c$	1/137.036

$\hbar c = 197.329$ MeV fm, $e^2/4\pi\varepsilon_0 = 1.43998$ MeV fm
1 MeV $= 1.60219 \times 10^{-13}$ J

$1\ \mathrm{fm} = 10^{-15}\ \mathrm{m}, \quad 1\ \mathrm{barn} = 10^{-28}\ \mathrm{m}^2 = 10^2\ \mathrm{fm}^2$

(Source: Review of particle properties (1984), *Rev. Mod. Phys.* **56**, No. 2, Part II)

Notation

$\mathbf{r}$, $\mathbf{k}$, etc., denote vectors (x, y, z), (k_x, k_y, k_z), and $r = |\mathbf{r}|$, $k = |\mathbf{k}|$,

$\mathrm{d}^3\mathbf{r} = \mathrm{d}x\,\mathrm{d}y\,\mathrm{d}z, \quad \mathrm{d}^3\mathbf{k} = \mathrm{d}k_x\,\mathrm{d}k_y\,\mathrm{d}k_z.$

$$\nabla^2 = \frac{\partial^2}{\partial x^2} + \frac{\partial^2}{\partial y^2} + \frac{\partial^2}{\partial z^2} = \frac{1}{r}\frac{\partial^2}{\partial r^2} r + \frac{1}{r^2 \sin\theta}\frac{\partial}{\partial\theta}\sin\theta\frac{\partial}{\partial\theta} + \frac{1}{r^2\sin^2\theta}\frac{\partial^2}{\partial\phi^2},$$

$\mathrm{d}\Omega = \sin\theta\,\mathrm{d}\theta\,\mathrm{d}\phi$ denotes an infinitesimal element of solid angle.

Glossary of some important symbols

A	nuclear mass number $(=N+Z)$
$\mathbf{A}(\mathbf{r}, t)$	electromagnetic vector potential
a	§4.1 nuclear surface width; §4.5 bulk binding coefficient
$B(Z, N)$	binding energy of nucleus
$\mathbf{B}(\mathbf{r}, t)$	magnetic field
b	§4.5 surface tension coefficient; §13.1 impact parameter
$\mathbf{E}(\mathbf{r}, t)$	electric field
E	energy; E_n, E_p neutron energy, proton energy; E_n^F, E_p^F neutron, proton Fermi energy, measured from the bottom of the shell-model neutron potential well; E_G §8.3
$F(Z, E_e)$	§12.3 Coulomb correction factor in β-decay
$f(Z, E_0)$	§12.3 kinematic factor in total β-decay rate
G	§6.2 exponent in the tunnelling formula
g	§D.2 statistical factor in Breit–Wigner formula
g_L, g_s	§5.6 orbital and intrinsic magnetic moment coefficients
g_A, g_V, g_L	§12.6 axial, vector, lepton coupling constants
$\mathcal{G}(r_s/r_c)$	§6.2 tunnelling integral
$\mathbf{J}$	§C.3 total angular momentum operator
j	quantum number associated with $\mathbf{J}^2$
j_z	quantum number of J_z
$\mathbf{k}$	wave vector
k_F	value of $k=\lvert\mathbf{k}\rvert$ at the Fermi energy
$\mathbf{L}$	§C.1 orbital angular momentum operator

l quantum number associated with $\mathbf{L}^2$; Chapter 9, Chapter 13 mean free path
m quantum number of L_z; reduced mass
m_s quantum number of s_z
m_α mass of α-particle; m_a, m_{nuc} mass of atom, nucleus
N number of neutrons in nucleus
$n(E)$ density of states
$\mathcal{N}(E)$ integrated density of states
$\mathbf{P}$ momentum
Q §5.7 nuclear electric quadrupole moment; §6.1 kinetic energy release in nuclear reaction
q §9.4 fission probability
R §4.3 nuclear radius; §12.3 reaction rate
r_s, r_c §6.2 potential barrier parameters
$S_n(N, Z)$ §5.2 neutron separation energy
$S(E)$ §8.3 parameter of nuclear reaction cross-section for energies below the Coulomb barrier
$S_0(E), S_c(E)$ §12.3 electron (positron) energy spectrum without and with Coulomb correction
$\mathbf{s}$ §C.2 intrinsic angular momentum operator
s quantum number associated with $\mathbf{s}$; §4.5 symmetry energy coefficient
T kinetic energy
$T_{1/2}$ decay half life
t_{nuc} §5.2 nuclear time scale
t_p §9.4 prompt neutron life
U potential energy; $\bar{U}$ mean proton–neutron potential energy difference in nucleus
$u_l(r)$ radial wave-function
V normalisation volume; §3.3 $V(r)$ nucleon–nucleon potential
v velocity
Z atomic number (number of protons in nucleus)
Γ, Γ_i width, partial width, of an excited state
γ §13.1 relativistic factor $(1 - v^2/c^2)^{-\frac{1}{2}}$
δ §2.4 coefficient of pairing energy
ε_0 permittivity of free space
ε_F §11.1 Fermi energy of electron gas
$\boldsymbol{\mu}$ §5.5 magnetic dipole operator
μ_n, μ_p neutron, proton magnetic moment
μ §5.5 magnetic dipole moment; §11.1 stellar mass per electron; §13.3 photon linear attenuation coefficient
μ_0 permeability of free space
ν, dν §9.3 mean number of prompt neutrons, delayed neutrons, per fission
ρ §2.1 electric charge density; §13.2 mass density

ρ_{ch} §4.1 electric charge density in units of e

ρ_0 §4.3 nucleon *number* density in nuclear matter

$\rho_{nuc}, \rho_n, \rho_p$ *number* density of nuclei, neutrons, protons

$\boldsymbol{\sigma}$ §C.2 Pauli spin matrices

σ cross-section; σ_{tot}, σ_e, σ_f total, elastic, fission, cross-section

τ mean life; τ_{E1}, τ_{M1} electric, magnetic, dipole transition mean life; §7.4 $(\tau_i)^{-1}$ partial decay rate

Φ §3.4 meson field

ϕ electromagnetic scalar potential

$\psi(\mathbf{r})$ single particle wave-function

ψ_m §D.1 general wave-function

Ω_{SO}, Ω_T §3.3 angular terms in the nucleon–nucleon potential

ω angular frequency

1 Prologue

The world is made up of some 92 chemical elements, distinguished from each other by the electric charge Ze on the atomic nucleus. This charge is balanced by the charge carried by the Z electrons which together with the nucleus make up the neutral atom. The elements are also distinguished by their mass, more than 99% of which resides in the nucleus. Are there other distinguishing properties of nuclei? Have the nuclei been in existence since the beginning of time? Are there elements in the Universe which do not exist on Earth? What physical principles underlie the properties of nuclei? Why are their masses so closely correlated with their electric charges, and why are some nuclei radio-active? Radio-activity is used to man's benefit in medicine. Nuclear fission is exploited in power generation. But man's use of nuclear physics has also posed the terrible threat of nuclear weapons.

This book aims to set out the basic concepts which have been developed by nuclear physicists in their attempts to understand the nucleus. Besides satisfying our appetite for knowledge, these concepts must be understood if we are to make an informed judgment on the benefits and problems of nuclear technology.

After the discovery of the neutron by Chadwick in 1932, it was accepted that a nucleus of atomic number Z was made up of Z protons and some number N of neutrons. The proton and neutron were then thought to be elementary particles, although it is now clear that they are not but rather are themselves structured entities. We shall also see that in addition to neutrons and protons several other particles play an important, if indirect, role in the physics of nuclei. In this and the following two chapters, to provide a background to our subsequent study of the nucleus, we shall describe the elementary particles of nature, and their interactions, as they are at present understood.

1.1 Fermions and bosons

Elementary particles are classified as either *fermions* or *bosons.* Fermions are particles which satisfy the Pauli exclusion principle: if an assembly of identical fermions is described in terms of single-particle wave-functions, then no two fermions can have the same wave-function. For example, electrons are fermions. This rule explains the shell structure of atoms and hence underlies the whole of chemistry. Fermions are so called because they obey the Fermi–Dirac statistics of statistical mechanics.

Bosons are particles which obey Bose–Einstein statistics, and are characterised by the property that *any* number of particles may be assigned the same single-particle wave-function. Thus, in the case of bosons, coherent waves of macroscopic amplitude can be constructed, and such waves may to a good approximation be described classically. For example, photons are bosons and the corresponding classical field is the familiar electromagnetic field **E** and **B**, which satisfies Maxwell's equations.

At a more fundamental level, these properties are a consequence of the possible symmetries of the wave-function of a system of identical particles when the coordinates of any two particles are interchanged. In the case of fermions, the wave-function changes sign; it is completely antisymmetric. In the case of bosons the wave-function is unchanged; it is completely symmetric.

There is also an observed relation between the intrinsic angular momentum, or spin, of a particle and its statistics. The intrinsic spin $\mathbf{s}$ is quantised, with spin quantum number s (see Appendix C). For a fermion, s takes one of the values $\frac{1}{2}, \frac{3}{2}, \frac{5}{2}, \ldots$; for a boson, s takes one of the values $0, 1, 2, \ldots$. A theoretical explanation of this relationship can be given within the framework of relativistic quantum field theory.

1.2 The particle physicist's picture of nature

Elementary particle physics describes the world in terms of elementary fermions, interacting through fields of which they are sources. The particles associated with the interaction fields are bosons. To take the most familiar example, an electron is an elementary fermion; it carries electric charge $-e$ and this charge produces an electromagnetic field **E**, **B**, which exerts forces on other charged particles. The electromagnetic field, quantised according to the rules of quantum mechanics, corresponds to an assembly of *photons*, which are bosons. Indeed, Bose–Einstein statistics were first applied to photons.

Four types of interaction field may be distinguished in nature (see Table 1.1). All of these interactions are relevant to nuclear physics, though the gravitational field becomes important only in densely aggregated matter, such as stars. Gravitational forces act on all particles and are important for the physics on the large scale of macroscopic bodies. On the small scale of most terrestrial atomic and nuclear physics, gravitational forces are insignificant and except in Chapter 10 and Chapter 11 we shall ignore them.

Nature provides an even greater diversity of elementary fermions than of bosons. It is convenient to divide the fermions into two classes: *leptons*, which are not sources of the strong fields and hence do not participate in the strong interaction; and *hadrons*, which take part in all interactions. The leptons and their interactions are described in Chapter 2. The elementary hadrons, and the proton and the neutron, form the subject matter of Chapter 3.

1.3 Conservation laws and symmetries; parity

The total energy of an isolated system is constant in time. So also are its linear momentum and angular momentum. These conservation laws are derivable from Newton's laws of motion and Maxwell's equations, or from the laws of quantum mechanics, but they can also, at a deeper level, be regarded as consequences of 'symmetries' of space and time. Thus the law of conservation of linear momentum follows from the homogeneity of space, the law of conservation of angular momentum from the isotropy of space; it does not matter where we place the origin of our coordinate axes, or in which direction they are oriented.

These conservation laws are as significant in nuclear physics as elsewhere, but there is another symmetry and conservation law which is of

Table 1.1. *Types of interaction field*

Interaction field	Boson	Spin
Gravitational field	'Gravitons' postulated	2
Weak field	W^+, W^-, Z particles	1
Electromagnetic field	Photons	1
Strong field	'Gluons' postulated	1

particular importance in quantum systems such as the nucleus: reflection symmetry and parity. By reflection symmetry we mean reflection about the origin, $\mathbf{r} \rightarrow \mathbf{r}' = -\mathbf{r}$. A single-particle wave-function $\psi(\mathbf{r})$ is said to have parity $+1$ if it is even under reflection, i.e.

$$\psi(-\mathbf{r}) = \psi(\mathbf{r}),$$

and parity -1 if it is odd under reflection, i.e.

$$\psi(-\mathbf{r}) = -\psi(\mathbf{r}).$$

More generally, a many-particle wave-function has parity $+1$ if it is even under reflection of all the particle coordinates, and parity -1 if it is odd under reflection.

Parity is an important concept because the laws of the electromagnetic and of the strong interaction are of exactly the same form if written with respect to a reflected left-handed coordinate system $(0x', 0y', 0z')$ as they are in the standard right-handed system $(0x, 0y, 0z)$ (Fig. 1.1). We shall see in Chapter 2 that this is not true of the weak interaction. Nevertheless, for many properties of atomic and nuclear systems the weak interaction is unimportant and wave-functions for such systems can be chosen to have a definite parity which does not change with time.

1.4 Units

Every branch of physics tends to find certain units particularly congenial. In nuclear physics, the size of the nucleus makes 10^{-15} m $= 1$ fm (femtometre) convenient as a unit of length, usually called a *fermi*. However,

1.1 The point P at $\mathbf{r}$ with coordinates (x, y, z) has coordinates $(-x, -y, -z)$ in the primed, reflected coordinate axes. $(0x', 0y', 0z')$ make up a *left-handed* set of axes.

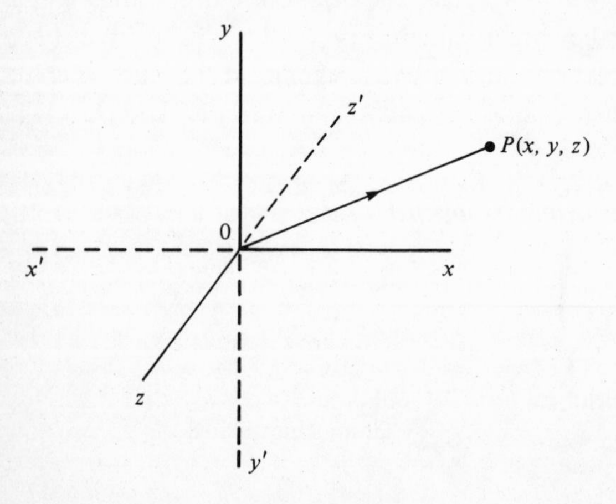

nuclear cross-sections, which have the dimensions of area, are measured in *barns*; $1\ \mathrm{b} = 10^{-28}\ \mathrm{m}^2 = 100\ \mathrm{fm}^2$. Energies of interest are usually of the order of MeV. Since mc^2 has the dimensions of energy, it is convenient to quote masses in units of MeV/c^2.

For order-of-magnitude calculations, the masses m_e and m_p of the electron and proton may be taken as

$$m_e \approx 0.5\ \mathrm{MeV}/c^2$$

$$m_p \approx 938\ \mathrm{MeV}/c^2$$

and it is useful to remember that

$$\hbar c \approx 197\ \mathrm{MeV\ fm}, \quad e^2/4\pi\varepsilon_0 \approx 1.44\ \mathrm{MeV\ fm},$$

$$e^2/4\pi\varepsilon_0 \hbar c \approx 1/137, \quad c \approx 3 \times 10^{23}\ \mathrm{fm\ s}^{-1}.$$

The student will perhaps be surprised to find how easily many expressions in nuclear physics can be evaluated using these quantities.

Problems

1.1 Show that the ratio of the gravitational potential energy to the Coulomb potential energy between two electrons is $\approx 2.4 \times 10^{-43}$.

1.2(*a*) Show that in polar coordinates (r, θ, ϕ) the reflection

$$\mathbf{r} \to \mathbf{r}' = -\mathbf{r} \quad \text{is equivalent to} \quad r \to r' = r$$

$$\theta \to \theta' = \pi - \theta, \quad \phi \to \phi' = \phi + \pi.$$

(*b*) What are the parities of the following electron states of the hydrogen atom:

(*i*) $$\psi_{100} = \frac{1}{\sqrt{\pi}}\left(\frac{1}{a_0}\right)^{\frac{3}{2}} e^{-r/a_0},$$

(*ii*) $$\psi_{210} = \frac{1}{4\sqrt{(2\pi)}}\left(\frac{1}{a_0}\right)^{\frac{3}{2}} \frac{r}{a_0} e^{-r/2a_0} \cos\theta,$$

(*iii*) $$\psi_{21-1} = \frac{1}{8\sqrt{\pi}}\left(\frac{1}{a_0}\right)^{\frac{3}{2}} \frac{r}{a_0} e^{-r/2a_0} \sin\theta\, e^{-i\phi}?$$

($a_0 = (4\pi\varepsilon_0)\hbar^2/m_e e^2$ is the Bohr radius.)

1.3(*a*) Show that the wavelength of a photon of energy 1 MeV is ≈ 1240 fm.

(*b*) The electrostatic self-energy of a uniformly charged sphere of total charge e, radius R, is $U = (3/5)e^2/(4\pi\varepsilon_0 R)$. Show that if $R = 1$ fm, $U = 0.86$ MeV.

2

Leptons and the electromagnetic and weak interactions

2.1 The electromagnetic interaction

The electromagnetic field is most conveniently described by a vector potential **A** and a scalar potential ϕ. For simplicity, we consider only the potential $\phi(\mathbf{r}, t)$. Using Maxwell's equations, this may be chosen to satisfy the wave equation

$$\nabla^2\phi - \frac{1}{c^2}\frac{\partial^2\phi}{\partial t^2} = -\frac{\rho(\mathbf{r}, t)}{\varepsilon_0}. \tag{2.1}$$

Here $\rho(\mathbf{r}, t)$ is the electric charge density due to the charged particles, which in atomic and nuclear physics will usually be electrons and protons, and c is the velocity of light.

In regions where $\rho = 0$, equation (2.1) has solutions in the form of propagating waves; for example, the plane wave

$$\phi(\mathbf{r}, t) = (\text{constant})\, e^{i(\mathbf{k}\cdot\mathbf{r} - \omega t)}. \tag{2.2}$$

This satisfies

$$\nabla^2\phi - \frac{1}{c^2}\frac{\partial^2\phi}{\partial t^2} = 0 \tag{2.3}$$

provided

$$\omega^2 = c^2k^2. \tag{2.4}$$

The wave velocity is therefore c, as we should expect. In quantum theory, unlike classical theory, the total energy and momentum of the wave are quantised, and can only be integer multiples of the basic quantum of energy and momentum given by the de Broglie relations:

$$E=\hbar\omega, \quad \mathbf{p}=\hbar\mathbf{k}. \tag{2.5}$$

Such a quantum of radiation is called a *photon*. A macroscopic wave can be considered to be an assembly of photons, and we can regard photons as particles, each carrying energy E and momentum $\mathbf{p}$.

Using (2.4) and (2.5), E and $\mathbf{p}$ are related by

$$E^2=p^2c^2. \tag{2.6}$$

For a particle of mass m, the Einstein equation gives

$$E^2=p^2c^2+m^2c^4.$$

We therefore infer that the photon is a particle having zero mass.

A second important type of solution of (2.1) exists when charged particles are present. If these are moving slowly compared with the velocity of light, so that the term $\partial^2\phi/(c^2\,\partial t^2)$ can be neglected, the solution is approximately the Coulomb potential of the charge distribution. For a particle with charge density ρ_1, we can take

$$\phi(\mathbf{r},t)\approx\frac{1}{4\pi\varepsilon_0}\int\frac{\rho_1(\mathbf{r}',t)}{|\mathbf{r}-\mathbf{r}'|}\,\mathrm{d}^3\mathbf{r}'. \tag{2.7}$$

Another charged particle with charge density ρ_2 will have a potential energy given by

$$\begin{aligned}U_{12}&=\int\rho_2(\mathbf{r},t)\phi(\mathbf{r},t)\,\mathrm{d}^3r\\&=\frac{1}{4\pi\varepsilon_0}\int\frac{\rho_1(\mathbf{r}',t)\rho_2(\mathbf{r},t)}{|\mathbf{r}-\mathbf{r}'|}\,\mathrm{d}^3\mathbf{r}\,\mathrm{d}^3\mathbf{r}'.\end{aligned} \tag{2.8}$$

Electric potential energy is basically responsible for the binding of electrons in atoms and molecules. We shall see that, in nuclear physics, it is responsible for the instability of heavy nuclei. If magnetic effects due to the motion of the charges are included, equation (2.8) is modified to

$$U_{12}=\frac{1}{4\pi\varepsilon_0}\int\frac{\rho_1'\rho_2+(1/c^2)\mathbf{j}_1'\cdot\mathbf{j}_2}{|\mathbf{r}-\mathbf{r}'|}\,\mathrm{d}^3\mathbf{r}\,\mathrm{d}^3\mathbf{r}', \tag{2.9}$$

where $\mathbf{j}=\rho\mathbf{v}$ is the current associated with the charge distribution which has velocity $\mathbf{v}(\mathbf{r})$. Thus this magnetic contribution to the energy is of relative order v^2/c^2.

The electromagnetic interaction also gives rise to the scattering of charged particles. For example, if ρ_1 and ρ_2 represent the charge distributions of two electrons approaching each other the interaction gives a mutual repulsion which leads to a transfer of momentum between the particles. The process can be represented by a diagram such as Fig. 2.1. In quantum electrodynamics, these diagrams, invented by Feynman, have a precise technical interpretation in the theory. We shall use them only to help visualise the physics involved. The scattering of the two electrons may be thought of as caused by the emission of a 'virtual' photon by one electron and its absorption by the other electron. In a virtual process the photon does not actually appear to an observer, though it appears in the mathematical formalism that describes the process.

2.2 The weak interaction

There are three *weak interaction* fields associated with the W^+, W^- and Z particles. Each one, like the electromagnetic field, is described by a vector and a scalar potential. However, the bosons associated with the weak fields all have mass, and the W^- and W^+ bosons are electrically charged. The Z boson is neutral, and most similar to the photon, but it has a mass

$$M_Z = (92.9 \pm 1.6)\ \mathrm{GeV}/c^2 \sim 100 \text{ proton masses},$$

which is very large by nuclear physics standards.

The interactions between leptons and the electromagnetic and weak fields were combined into a unified 'electro-weak' theory by Weinberg and by Salam. The existence of the Z and $W^\pm$ bosons was predicted by the theory, and the theory also suggested values for their masses. These predictions were confirmed by experiments at CERN in 1983.

The wave equation satisfied by the scalar potential ϕ_Z associated with the

2.1 The scattering of two electrons of momenta $\hbar\mathbf{k}$, $\hbar\mathbf{k}'$ by the exchange of a virtual photon carrying momentum $\hbar\mathbf{q}$. Time runs from left to right in these diagrams. (In principle, the exchange of a Z particle (§2.2) also contributes to electron–electron scattering, but the very short range and weakness of the weak interaction makes this contribution almost completely negligible: the electrons are in any case kept apart by the Coulomb repulsion induced by the photon exchange.)

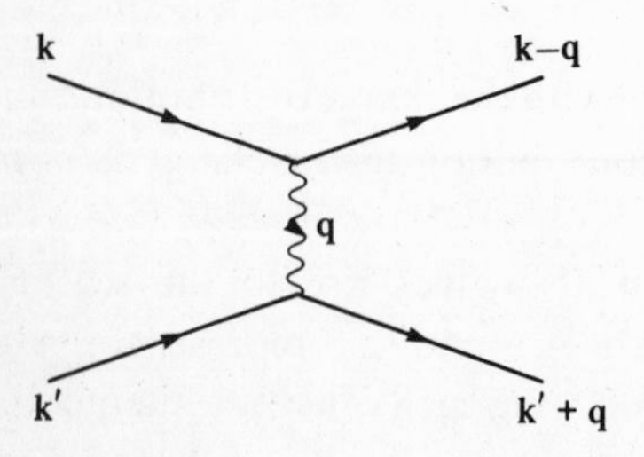

Z boson is a generalisation of (2.1) and includes a term involving M_Z:

$$\left[\nabla^2 - \frac{1}{c^2}\frac{\partial^2}{\partial t^2} - \left(\frac{M_Z c}{\hbar}\right)^2\right]\phi_Z(\mathbf{r}, t) = -\frac{\rho_Z(\mathbf{r}, t)}{\varepsilon_0}, \tag{2.10}$$

where ρ_Z is the neutral weak-charge density. There is a close, but not exact, analogy between weak-charge density and electric-charge density, and particles carry weak charge somewhat as they carry electric charge.

In free space where $\rho_Z = 0$ there exist plane wave solutions of (2.10),

$$\phi_Z(\mathbf{r}, t) = (\text{constant})\ e^{i(\mathbf{k}\cdot\mathbf{r} - \omega t)},$$

but now to satisfy the wave equation we require

$$\omega^2 = c^2 k^2 + c^2 (M_Z c/\hbar)^2,$$

and with the de Broglie relations (2.5) for the field quanta we obtain the Einstein energy–momentum relation for the Z boson:

$$E^2 = p^2 c^2 + M_Z^2 c^4.$$

The static solution of (2.10) which corresponds to a point unit weak charge at the origin is

$$\phi_Z(r) = \frac{1}{4\pi\varepsilon_0}\frac{e^{-\kappa r}}{r}, \quad \text{writing } \kappa = \frac{M_Z c}{\hbar}. \tag{2.11}$$

At points away from the origin where $\nabla^2\phi_Z - \kappa^2\phi_Z = 0$, this satisfies equation (2.10), as may be easily checked by substitution, using the formula $\nabla^2\phi_Z = (1/r)\,d^2(r\phi_Z)/dr^2$. Close to the origin the solution (2.11) behaves like the corresponding Coulomb potential $1/(4\pi\varepsilon_0 r)$ of a unit point electric charge, and hence has the correct point source behaviour. The generalisation of (2.11) to a distribution of weak charge gives the quasi-static solution (cf. (2.7))

$$\phi_Z(\mathbf{r}, t) \approx \frac{1}{4\pi\varepsilon_0}\int \frac{\rho_Z(\mathbf{r}', t)\, e^{-\kappa|\mathbf{r}-\mathbf{r}'|}}{|\mathbf{r}-\mathbf{r}'|}\, d^3\mathbf{r}'. \tag{2.12}$$

The exponential factor in the integral effectively vanishes for $|\mathbf{r}-\mathbf{r}'|$ greater than a few times $\kappa^{-1} = \hbar/M_Z c$ and

$$\hbar/M_Z c \approx 2 \times 10^{-3}\ \text{fm}.$$

This is a very small distance in the context of nuclear physics: by the uncertainty principle, low momentum sources must be spread over distances much greater than this. Hence in the integral in (2.12) the factor ρ_Z is slowly varying over the range of the exponential and may be taken outside the integral (which is then elementary):

$$\begin{aligned}\phi_Z(\mathbf{r}, t) &\simeq \frac{1}{4\pi\varepsilon_0} \rho_Z(\mathbf{r}, t) \int \frac{e^{-\kappa|\mathbf{r}-\mathbf{r}'|}}{|\mathbf{r}-\mathbf{r}'|} \mathrm{d}^3\mathbf{r}' \\ &= \frac{1}{4\pi\varepsilon_0} \rho_Z(\mathbf{r}, t) \int_0^\infty \frac{e^{-\kappa R}}{R} 4\pi R^2 \mathrm{d}R \\ &= \frac{1}{\varepsilon_0}\left(\frac{\hbar}{M_Z c}\right)^2 \rho_Z(\mathbf{r}, t).\end{aligned} \tag{2.13}$$

The potential energy between two particles associated with the scalar field ϕ_Z is, by analogy with (2.8),

$$\begin{aligned}U_{12}^Z &= \int \rho_{Z2}(\mathbf{r}, t)\phi_{Z1}(\mathbf{r}, t)\, \mathrm{d}^3\mathbf{r} \\ &\approx \frac{1}{\varepsilon_0}\left(\frac{\hbar}{M_Z c}\right)^2 \int \rho_{Z1}(\mathbf{r}, t)\rho_{Z2}(\mathbf{r}, t)\, \mathrm{d}^3\mathbf{r},\end{aligned}$$

and there is also a contribution from the vector part of the field, analogous to the magnetic contribution in (2.9), of the form

$$\frac{1}{\varepsilon_0 c^2}\left(\frac{\hbar}{M_Z c}\right)^2 \int \mathbf{j}_{Z1}(\mathbf{r}, t)\cdot\mathbf{j}_{Z2}(\mathbf{r}, t)\, \mathrm{d}^3\mathbf{r},$$

where $\mathbf{j}_Z$ is the weak-current density.

The physical consequences of these expressions are quite different from the electromagnetic interaction. U_{12}^Z is very much suppressed by the large mass factor in the denominator, and it is this which largely accounts for the 'weakness' of the weak interaction. Also the interaction at low energies appears as a 'contact interaction', effectively having zero range.

The electrically charged W^+ and W^- boson fields give rise to the most important weak interactions, and in particular to β-decay. They obey equations similar to those of the Z field, but the masses of the associated particles are somewhat smaller;

$$M_{W^+} = M_{W^-} = (80.8 \pm 2.7)\ \mathrm{GeV}/c^2.$$

2.3 Mean life and half life

Not all particles are stable: some, for example the $W^\pm$ and Z bosons, have only a transient existence. Suppose that an unstable particle exists at some instant $t = 0$; its *mean life* is the mean time it exists in isolation, before it undergoes radio-active decay. If we denote by $P(t)$ the probability that the particle survives for a time t, and make the basic assumption that the particle has a *constant* probability $1/\tau$ per unit time of decaying, then

$$P(t+\mathrm{d}t)=P(t)(1-\mathrm{d}t/\tau),$$

since $(1-\mathrm{d}t/\tau)$ is the probability it survives the time interval $\mathrm{d}t$. Hence

$$\frac{1}{P}\frac{\mathrm{d}P}{\mathrm{d}t}=-\frac{1}{\tau},$$

and integrating,

$$P(t)=P(0)e^{-t/\tau}.$$

Since $P(0)=1$ we have

$$P(t)=e^{-t/\tau}. \tag{2.14}$$

Equation (2.14) is the familiar exponential-decay law for unstable particles. It is well verified experimentally.

The probability that the particle decays between times $t, t+\mathrm{d}t$ is clearly $P(t)\times(\mathrm{d}t/\tau)$, so that the mean life is

$$\int_0^\infty tP(t)(\mathrm{d}t/\tau)=\int_0^\infty te^{-t/\tau}\,\mathrm{d}t/\tau=\tau.$$

The 'half life' $T_{\frac{1}{2}}$ is the time at which there is a 50% probability that the particle has decayed, i.e.

$$P(T_{\frac{1}{2}})=e^{-T_{\frac{1}{2}}/\tau}=\tfrac{1}{2}$$

Hence

$$T_{\frac{1}{2}}=\tau\log 2=0.693\tau.$$

In this book we have preferred to quote mean lives rather than half lives. We refer to $(1/\tau)$ as the *decay rate*.

2.4 Leptons

Leptons are spin $\frac{1}{2}$ fermions which interact through the electromagnetic and weak interactions, but not through the strong interaction. The known leptons are listed in Table 2.1.

Table 2.1. *Known leptons*

	Mass (MeV/c^2)	Mean life (s)	Charge
Electron e^-	0.5110	∞	$-e$
Electron neutrino ν_e	$<46\times10^{-6}$	∞	0
Muon μ^-	105.659	2.197×10^{-6}	$-e$
Muon neutrino ν_μ	<0.5	∞?	0
Tau τ^-	1784	$(3.4\pm0.5)\times10^{-13}$	$-e$
Tau neutrino ν_τ	<164	∞?	0

The electrically charged leptons all have magnetic moments of magnitude $\approx -e\hbar/2$ (mass) aligned with their spins.

Of these charged leptons, only the familiar electron is stable. Electrons are structureless particles that are described by the Dirac relativistic wave-equation. This equation explains the spin and magnetic moment of the electron, and has the remarkable feature that it predicts the existence of anti-particles: these are particles of the same mass and spin, but of opposite charge and magnetic moment to the particle. The anti-particle of the electron is called the *positron*. Positrons were identified experimentally by Anderson in 1932 soon after their theoretical prediction.

Since leptons do not interact with the strong interaction field, electrons and positrons interact principally through the electromagnetic field. A positron will eventually annihilate with an electron, usually to produce two or three photons, so that all the lepton energy appears as electromagnetic radiation. We write these processes as

$$\mathrm{e}^- + \mathrm{e}^+ \rightarrow 2\gamma$$
$$\mathrm{e}^- + \mathrm{e}^+ \rightarrow 3\gamma.$$

The converse processes of *pair-production* by photons are also possible, and pair-production from a single photon is possible provided another (charged) particle is present to take up momentum. Quantum electrodynamics, based on the Dirac and Maxwell equations, describes all processes involving electrons, positrons and photons to a high degree of accuracy.

It is a curious fact that nature provides us also with the electrically charged *muon* μ^- and *tau* τ^- and their anti-particles the μ^+ and τ^+. Apart from their greater masses and finite lifetimes, muons and taus seem to be just copies of the electron, and like the electron they are accurately described by Dirac equations. We shall see that the μ^- can be used as a probe of nuclear charge density, but otherwise neither the muons nor the taus play any significant role in nuclear physics.

The remaining leptons are the *neutrinos* ν and their corresponding anti-neutrinos denoted by $\bar{\nu}$. All the experimental evidence is consistent with their mass being zero, so that, like photons, they move with the speed of light. However, neutrinos are fermions with spin $\frac{1}{2}$.

It is exceedingly difficult and expensive to carry out experiments with neutrinos, but there is evidence that the electron, muon and tau have different neutrinos, ν_e, ν_μ, ν_τ associated with them.

2.5 The instability of the heavy leptons: muon decay

The W^+ and W^- bosons lead to processes called β-decay, which neither photons nor Z bosons can induce. In this chapter we illustrate this with the example of the β-decay of the muon; in the next chapter we shall describe β-decay processes involving hadrons.

The muon decays to a muon neutrino, together with an electron and an electron anti-neutrino:

$$\mu^- \to \nu_\mu + e^- + \bar{\nu}_e.$$

The W fields play the mediating role in this decay through the two virtual processes illustrated in Fig. 2.2. Again, in a virtual process actual W bosons do not appear to an observer.

The W bosons can in principle produce any charged lepton and its anti-neutrino or an anti-lepton and its neutrino, but energy must be conserved overall. Hence in the case of muon decay the charged lepton must be an electron. A tau decay can produce a muon or an electron (and indeed it is sufficiently massive to decay alternatively to hadrons, as we shall see later).

It is of fundamental significance that electric charge is conserved at every stage of a decay. It is also believed to be true of all interactions that a single lepton can only change to another of the same type, and a lepton and an anti-lepton of the same type can only be created or destroyed together. There is thus a conservation law, the '*conservation of lepton number*' (anti-leptons being counted negatively), for each separate type of lepton.

2.6 Parity violation in muon decay

It is observed experimentally that in the decay of the negative muon, the electron momentum $\mathbf{p}_e$ is strongly biased to be in the direction opposite to that of the muon spin $\mathbf{s}_\mu$. To explain the implication of this

2.2 The decay $\mu^- \to \nu_\mu + e^- + \bar{\nu}_e$. In (*a*) the muon changes to its neutrino and a 'virtual' W^- boson, which then decays to the electron and the electron anti-neutrino. In (*b*) a 'virtual' W^+ is created from the vacuum with the electron and the electron anti-neutrino. The W^+ then transforms the muon into a muon neutrino.

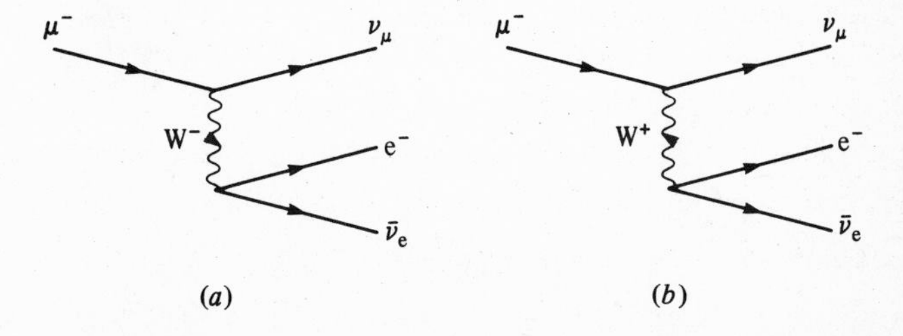

observation for parity violation, we must first point out that there are two types of vector.

Under the reflection in the origin (Fig. 1.1), the position vector $\mathbf{r}$ of a particle and its momentum $\mathbf{p}$ transform:

$$\mathbf{r} \to \mathbf{r}' = -\mathbf{r} \quad \text{and} \quad \mathbf{p} = m\frac{\mathrm{d}\mathbf{r}}{\mathrm{d}t} \to \mathbf{p}' = m\frac{-\mathrm{d}\mathbf{r}}{\mathrm{d}t} = -\mathbf{p}, \tag{2.15}$$

$\mathbf{r}$ and $\mathbf{p}$ are both *true vectors.*

The angular momentum $\mathbf{L} = \mathbf{r} \times \mathbf{p}$ has many of the attributes of a vector, but under reflection

$$\mathbf{L} \to \mathbf{L}' = (-\mathbf{r}) \times (-\mathbf{p}) = +\mathbf{L}.$$

Thus $\mathbf{L}$ does not have the reflection property (2.14) of the true vectors $\mathbf{r}$ and $\mathbf{p}$. It is called an *axial vector* or *pseudo-vector.* The intrinsic angular momentum $\mathbf{s}$ of a particle is likewise an axial vector.

Returning to muon decay, in the reflected coordinate system, $\mathbf{p}_e \to -\mathbf{p}_e$, $\mathbf{s}_\mu \to +\mathbf{s}_\mu$, so that the momentum would be said to be biased in the same direction as the muon spin! It appears that the equations of the theory are only valid in the original right-handed frame, and would have to be rewritten to hold in the left-handed reflected frame. Thus the laws are not invariant under reflection and hence parity is not conserved in muon decay. More generally, parity is not conserved in any process involving the weak interaction fields.

The inequivalence of right-handedness and left-handedness is most extreme in the case of neutrinos. Neutrinos produced in a weak interaction process are always 'left-handed', with their spin anti-parallel to their direction of motion, and anti-neutrinos are always 'right-handed' (Fig. 2.3). There is no evidence that right-handed neutrinos (or left-handed anti-neutrinos) exist at all.

The breakdown of parity conservation may be expressed slightly differently. The reflection in the origin $\mathbf{r} \to \mathbf{r}' = -\mathbf{r}$ is easily seen to be equivalent to mirror reflection in a plane, followed by a rotation through π

2.3 The relation between momentum $\mathbf{p}$ and spin for a neutrino ν and an anti-neutrino $\bar{\nu}$.

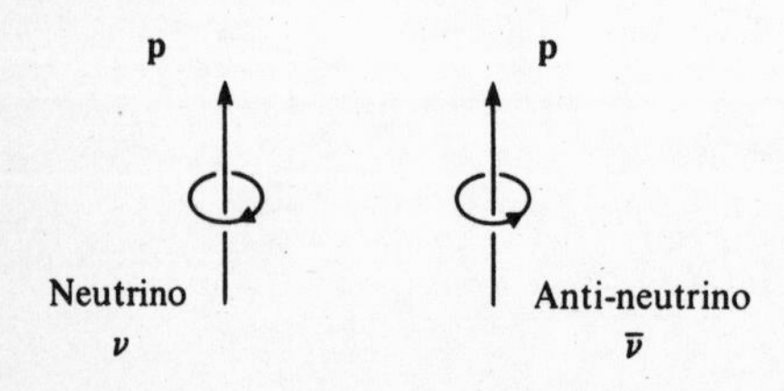

about an axis perpendicular to that plane (e.g. the xy-plane and the z-axis, cf. Problem 1.2). There is no evidence that the laws of physics break down under rotations, so the breakdown is in the mirror reflection: the assumption that the mirror image of a physical process is also a possible physical process is wrong, in so far as the weak interaction is involved.

Problems

2.1 Plane wave solutions of the relativistic wave-equation for a free particle of mass m are of the form

$$\psi(\mathbf{r}, t) = (\text{constant})e^{i(\mathbf{k}\cdot\mathbf{r}-\omega t)}$$

where

$$\omega^2 = c^2k^2 + (m^2c^4/\hbar^2).$$

Show that the group velocity of a wave-packet representing a particle of total energy $E=\hbar\omega$ is the same as the velocity of a relativistic classical particle having the same total energy.

2.2 The weak charge density of an electron bound in an atom has a similar magnitude to the electric charge density and has, similarly, a probability distribution over the atomic dimensions of the electron's wave-function. Show that the ratio of the weak interaction energy to the electrostatic interaction energy between two electrons bound in an atom is of order of magnitude $4\pi(\hbar/(a_0M_Zc))^2 \sim 10^{-15}$, where a_0 is the Bohr radius. (Compare this result with Problem 1.1.)

2.3 An electron–positron pair bound by their Coulomb attraction is called *positronium*. Show that when positronium decays from rest to two photons, the photons have equal energy.

2.4 Use energy and momentum conservation to show that pair creation by a single photon, $\gamma \rightarrow e^+ + e^-$, is impossible in free space.

2.5 Show that a muon in free space with a kinetic energy of 1 MeV will travel a mean distance of about 90 m before it decays.

2.6 An electron and a μ^+ bound by their Coloumb attraction is called *muonium*. Which of the following decays can occur:

(*a*) $(\mu^+ e^-) \rightarrow \gamma + \gamma$

(*b*) $(\mu^+ e^-) \rightarrow \nu_e + \bar{\nu}_\mu$

(*c*) $(\mu^+ e^-) \rightarrow e^+ + e^- + \nu_e + \bar{\nu}_\mu$?

2.7 The masses of the electron and neutrinos from a muon decay are negligible compared with the muon mass. Show that if the muon decays from rest and the kinetic energy released is divided equally between the final leptons then the angle between the paths of any two of them is approximately 120°.

2.8 Starting from the Coulomb law and the Biot–Savart law, show that the electric field **E** is a true vector field, but that the magnetic field **B** is an axial vector field.

3
Nucleons and the strong interaction

We turn now to the hadrons, the particles which interact by the *strong interaction*, as well as by the weak and electromagnetic interactions. In particular we shall describe the *nucleons*, that is to say, the *proton* and the *neutron*, the forces between nucleons, and the effect of the weak interaction on the stability of nucleons.

3.1 Properties of the proton and the neutron

Nucleons, like leptons, are fermions with spin $\frac{1}{2}$. The mass of the neutron is 0.14% greater than that of the proton:

$$\begin{aligned} m_{\mathrm{n}} &= 939.57\ \mathrm{MeV}/c^2, \\ m_{\mathrm{p}} &= 938.28\ \mathrm{MeV}/c^2. \end{aligned} \tag{3.1}$$

Thus the mass difference $m_{\mathrm{n}} - m_{\mathrm{p}} = 1.29\ \mathrm{MeV}/c^2$ (≈ 2 electron masses).

The neutron has no net electric charge. The proton has the opposite charge to the electron: protons are responsible for exactly cancelling the charge of the electrons in electrically neutral atoms.

The electric charge on a proton is not concentrated at a point, but is symmetrically distributed about the centre of the proton. By the experimental methods to be discussed in Chapter 4, the mean radius R_{p} of this charge distribution is found to be $R_{\mathrm{p}} \approx 0.8$ fm. An extended charge distribution is also found in the neutron, positive charge in the central region being cancelled by negative charge at greater distances. The matter distribution in nucleons also extends to a distance of about R_p.

Both the proton and the neutron have a magnetic dipole moment, aligned with their spin:

$$\begin{aligned} \mu_p &= 2.79284(e\hbar/2m_p), \\ \mu_n &= -1.91304(e\hbar/2m_p). \end{aligned} \tag{3.2}$$

Clearly neither magnetic moment is simply related to the value $e\hbar/[2(\text{nucleon mass})]$ expected from a simple Dirac equation, and this is a clear indication that the nucleons are not themselves fundamental particles.

Compelling evidence that the nucleons are the ground states of a composite system is given by data of which that in Fig. 3.1 is an example. This shows the cross-section for absorption of photons by protons and by deuterons (see § 3.3), as a function of photon energy up to 1300 MeV. The

3.1 The total photon cross-section for hadron production on protons (dashes) and deuterons (crosses). The difference between these cross-sections is approximately the cross-section on neutrons. (After Armstrong, T. A. *et al.* (1972), *Phys. Rev.* **D5**, 1640; *Nuc. Phys.* **B41**, 445.)

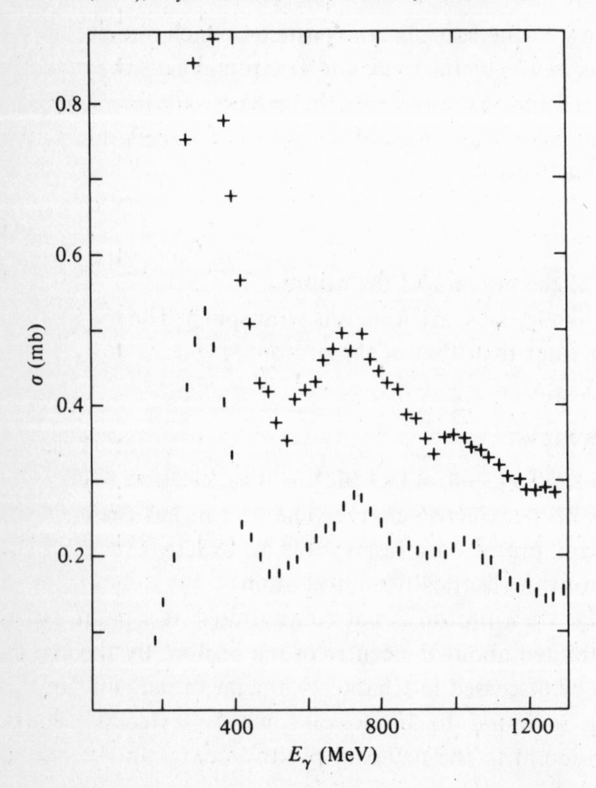

cross-sections vary rapidly with energy. A precise definition of cross-section is given in Appendix A, but for our immediate purpose it is sufficient to remark that the peaks are due to photons being preferentially absorbed to create an excited state when the photon energy matches the excitation energy of that state. Perhaps a more familiar example of photons being absorbed by a composite system is that of atomic absorption. Similar peaks in atomic absorption cross-sections, but at energies of a few electron volts, correspond to the excitation of the atom to higher energy states. The nucleon peaks have a similar interpretation, albeit on a very different energy scale. The first peak in the proton cross-section is at a photon energy of about 294 MeV, and corresponds to the formation of a state called the Δ^+. The Δ^+ is a fermion with mass of about $(938+294)$ MeV $\approx$ 1232 MeV; its spin has been determined to be $\frac{3}{2}$.

Data for the neutron show that it has a sequence of excited states of the same spins and almost identical energies as has the proton. The electrical energies associated with the charge distributions of the proton and neutron are of order of magnitude $e^2/(4\pi\varepsilon_0 R_p) \approx 2$ MeV (taking $R_p = 0.8$ fm), which is small compared with the nucleon self-energies and excitation energies. We shall see that, in all strong interactions, protons and neutrons behave in the same way to a good approximation. The near independence of the strong interaction on nucleon type is an important fact for our understanding of the properties of the nucleus.

3.2 The quark model of nucleons

Any composite system with spin $\frac{1}{2}$ must contain an odd number of fermion constituents. (An even number would give integral spin.) The highly successful quark model postulates that nucleons contain three fundamental fermions called *quarks*. We cannot here present the particle physics which establishes the validity of the quark model, but since particle physics does have implications for the concepts of nuclear physics we give – without attempting justification – some of the most relevant results.

As is the case with the elementary leptons, there are several types of quark, with a curious and so far unexplained mass hierarchy. For nucleons and nuclear physics only the two least-massive quarks are involved, the up quark u and the down quark d. The proton basically contains two up quarks and a down quark (uud) and the neutron two downs and one up (ddu). These quarks are bound by the fundamental strong interaction field, called by particle physicists the *gluon* field. The fact that the strong interactions of neutrons are almost the same as those of protons is explained by the gluon field having the same coupling to all quarks, independent of their type.

What are the properties of these quarks? They have mass, but the mass of a particle is generally determined by isolating it and measuring its acceleration in response to a known force. Because a single quark has never been isolated, this procedure has not been possible, and our knowledge of the quark masses is indirect. The consensus is that much of the nucleon mass resides in the gluon force fields that bind the quarks, and only a few MeV/c^2 need be assigned to the u and d quark masses. It is well established that the d quark is heavier than the u quark, since in all cases where two particles differ only in that a d quark is substituted for a u quark, the particle with the d quark is heavier. The principal example of this is the difference in mass between the neutron and proton. The mass, $\sim 2\ \mathrm{MeV}/c^2$, associated with the electrical energy of the charged proton is far greater than that associated with the (overall neutral) charge distribution of the neutron, so that one might expect the proton to be heavier. However, the extra d quark in the neutron more than compensates for this, and makes the neutron heavier than the proton.

The electric charges carried by quarks are well verified by measurements of the electromagnetic transitions between the nucleon ground states and excited states. The u has charge $\frac{2}{3}e$ and the d has charge $-\frac{1}{3}e$. Thus the proton (uud) has net charge e and the neutron (ddu) has net charge zero. Again, since a quark has never been isolated, the evidence for these assignments is all indirect.

The differences between neutrons and protons, other than their electric and weak charges, are due to the u–d mass difference. This has only a small effect on the basic strong interactions, so that the resulting strong interaction between nucleons is almost independent of nucleon type. This independence may be expressed mathematically by introducing the concept of 'isotopic spin symmetry', but for our purposes this elaboration is unnecessary.

3.3 The nucleon–nucleon interaction: the phenomenological description

We shall see in later chapters that the kinetic energies and potential energies of nucleons bound together in a nucleus are an order of magnitude smaller than the energies (~ 290 MeV) required to excite the quarks in an individual nucleon. It is, therefore, reasonable to regard a nucleus as an assembly of nucleons interacting with each other, but basically remaining in their ground states. To understand the physics of nuclei it is therefore important to be able to describe the interactions between nucleons. Since nucleons are composite particles, we can anticipate that their interactions with each other will not be simple. In fact they are rather complicated. Nevertheless, after 50 years of experimental and theoretical effort a great deal is known empirically about the forces between two nucleons, especially at the low energies relevant to nuclear physics.

The empirical approach is to construct a possible potential which incorporates our limited theoretical knowledge (which we shall discuss in § 3.4) and has adjustable features, mainly to do with the short-range part of the interaction. The Schrödinger equation for two nucleons interacting through this potential is then solved numerically and the adjustable features are varied to fit the experimental facts, namely the properties of the *deuteron* and the low-energy scattering data.

The deuteron is a neutron–proton bound state with:

$$\begin{aligned} &\text{binding energy} = 2.23\ \text{MeV}, \\ &\text{angular momentum } j = 1, \\ &\text{magnetic moment} = 0.8574(e\hbar/2m_\text{p}), \\ &\text{electric quadrupole moment} = 0.286\ \text{fm}^2. \end{aligned} \tag{3.3}$$

Neither proton–proton nor neutron–neutron bound states exist.

The scattering data provide much more information. Nucleons have spin $\frac{1}{2}$, which may be 'flipped' in the scattering. It can be shown that there are five independent differential cross-sections for spin-polarized proton–proton and proton–neutron scattering which can, in principle, be measured. Neutron–neutron cross-sections have never been measured directly because there are no targets of free neutrons.

As has been explained, the strong neutron–neutron interaction should be almost the same as the strong proton–proton interaction, and both these should be almost the same as the proton–neutron interaction for the same states of relative motion. However, we must remember here the Pauli exclusion principle: the neutron and proton can exist together in states which are not allowed for two protons or two neutrons. This is why the neutron and proton can have a bound state, whereas two protons or two neutrons do not bind, without any contradiction of the principle that the strong interaction is almost independent of nucleon type.

A large amount of careful and accurate data has been accumulated, and the most sophisticated and accurate empirical potential has been constructed by a group of scientists working in Paris. Two expressions are needed: one for the (anti-symmetric) states allowed for two protons or two neutrons, as well as a proton and a neutron, and one for symmetric states accessible only to the neutron–proton system. For both cases, when the spins of the two nucleons are coupled to give a total spin $S=0$ (see Appendix C) the nucleons only experience a central potential.

When the spins couple to $S=1$ there are four contributions to these potentials, which are then each of the form

$$V(r)=V_{C1}(r)+V_T(r)\Omega_T+V_{SO}(r)\Omega_{SO}+V_{SO2}(r)\Omega_{SO2},$$

where

$$\Omega_T=3\frac{(\boldsymbol{\sigma}_1\cdot\mathbf{r})(\boldsymbol{\sigma}_2\cdot\mathbf{r})}{r^2}-\boldsymbol{\sigma}_1\cdot\boldsymbol{\sigma}_2 \tag{3.4}$$

$$\hbar\Omega_{SO}=(\boldsymbol{\sigma}_1+\boldsymbol{\sigma}_2)\cdot\mathbf{L}$$

$$\hbar^2\Omega_{SO2}=(\boldsymbol{\sigma}_1\cdot\mathbf{L})(\boldsymbol{\sigma}_2\cdot\mathbf{L})+(\boldsymbol{\sigma}_2\cdot\mathbf{L})(\boldsymbol{\sigma}_1\cdot\mathbf{L}).$$

In these expressions $\boldsymbol{\sigma}(\hbar/2)$ is the nucleon spin operator, $\mathbf{L}$ is the orbital angular momentum operator of the nucleon pair, and the subscripts 1 and 2 refer to the two nucleons present.

V_{C1} is essentially an ordinary central potential. $V_T\Omega_T$ is called the tensor potential. It has the same angular structure as the potential between two magnetic dipoles and it is also interesting because it is the only part of the potential which mixes states of different orbital angular momentum. $V_{SO}\Omega_{SO}$ and $V_{SO2}\Omega_{SO2}$ give rise to different terms for the different couplings of spin and orbital angular momenta. Spin orbit coupling is well known in atomic physics, where it is due to magnetic interactions. However, these terms in the nuclear potential, which are of major importance, arise out of the strong interaction.

In Fig. 3.2 we show the four potentials that are most important at low energies of interaction (<100 MeV) and in particular are important for nucleons in nuclei.

The potential $V_{C0}(r)$ is appropriate for low-energy proton–proton and neutron–neutron interactions. The attractive tail is not, however, sufficiently deep to bind two nucleons. The potentials $V_{C1}(r)$, $V_{SO}(r)$ and $V_T(r)$ are responsible for binding the deuteron: note the deeply attractive part of $V_T(r)$, which is associated also with the large electric quadrupole moment of the deuteron.

The central potentials have the important feature of a repulsive core at ~ 0.8 fm, which stops nuclei collapsing. The attractive part of these potentials binds nucleons together in nuclei. The tensor potential is particularly important for binding the deuteron, but since it is zero on averaging over all directions it becomes less important in heavier nuclei. This last remark presupposes that the potential established for the interaction of two nucleons in isolation is relevant when many nucleons are interacting in an atomic nucleus. We shall discuss this assumption further in Chapter 4.

3.4 Mesons and the nucleon–nucleon interaction

Like all fermions, quarks have corresponding anti-particles. Anti-protons and anti-neutrons can exist, made up of anti-quarks, ($\bar{u}\bar{u}\bar{d}$) and ($\bar{d}\bar{d}\bar{u}$); the excited states of nucleons have images of identical mass but opposite charge in anti-quark matter. In fact the electromagnetic and strong interactions of anti-matter seem to be identical to those of matter. It is possible to contemplate the existence of stable anti-atoms, and macroscopic bodies, made up of anti-matter, but as electrons annihilate with positrons, so do nucleons annihilate with anti-nucleons; matter and anti-matter, though stable in isolation, cannot coexist. To study anti-particles we must create them in laboratories.

3.2 The most important components of the 'Paris potential'. (After Lacombe, M. *et al.* (1980), *Phys. Rev.* **C21**, 861.)

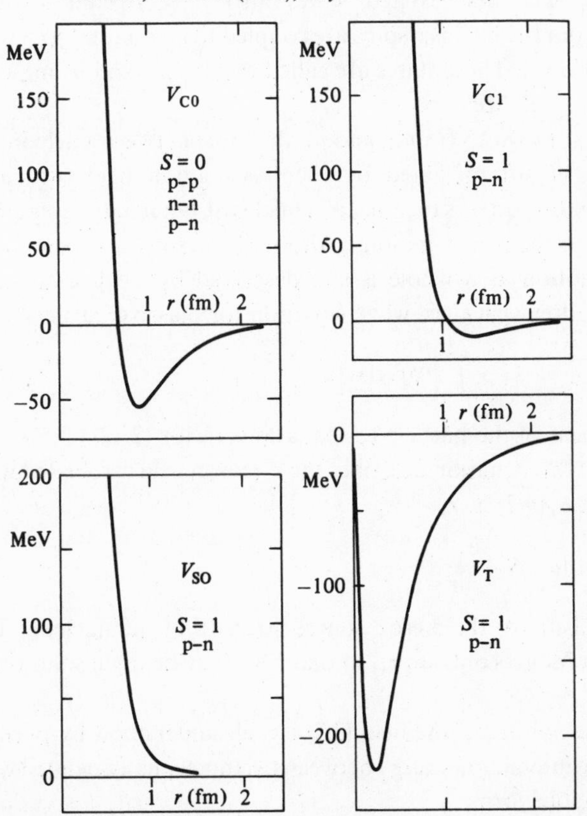

As well as binding three quarks or three anti-quarks together to make nucleons and anti-nucleons, the strong gluon field can bind a quark and an anti-quark together to form a short-lived particle called a *meson*. Like nucleons, such bound pairs have a sequence of excited states.

Of most importance for nuclear physics are the π-mesons. The electrically charged π^+ and π^- are made up of $(u\bar{d})$ and $(d\bar{u})$ pairs respectively, and the neutral π^0 is a superposition $(u\bar{u} - d\bar{d})/\sqrt{2}$ of quark anti-quark pairs. (The orthogonal combination $(u\bar{u} + d\bar{d})/\sqrt{2}$ belongs to a meson called the η.)

The masses of the π-mesons are:

$$\begin{aligned} &\text{mass of } \pi^+ = \text{mass of } \pi^- = 139.57 \text{ MeV}/c^2, \\ &\text{mass of } \pi^0 = 134.96 \text{ MeV}/c^2. \end{aligned} \tag{3.5}$$

(The η has mass ≈ 549 MeV$/c^2$.)

The quark anti-quark pairs in these mesons have orbital angular momentum zero and intrinsic spins coupled to give total angular momentum zero. The first excited states also have orbital angular momentum zero, but the intrinsic spins are coupled to give a total spin with quantum number $S = 1$. These states are called the ρ^+, ρ^- and ρ^0 mesons; they have masses ~ 750 MeV$/c^2$.

For reasons that are not yet understood, the force between nucleons at distances $\gtrsim 1$ fm is not mediated by the basic gluon field (which is responsible for holding quarks together in a nucleon), but it is apparent that it is due to the exchange of mesons. Although mesons are composite particles, their motion as a whole is still described by a wave-function $\Phi(\mathbf{r}, t)$, obeying in free space the wave-equation for massive particles:

$$\left[\nabla^2 - \frac{1}{c^2}\frac{\partial^2}{\partial t^2} - \left(\frac{mc}{\hbar}\right)^2\right]\Phi(\mathbf{r}, t) = 0, \tag{3.6}$$

where m is the mass of the particle (cf. equations (2.10)–(2.12).)

One solution of this equation describes the π-meson field associated with a nucleon of spin $\boldsymbol{\sigma}_1(\hbar/2)$ at $\mathbf{r}_1$:

$$\Phi(\mathbf{r}, t) = g_\pi(\boldsymbol{\sigma}_1 \cdot \boldsymbol{\nabla}_1)\frac{e^{-mc|\mathbf{r}-\mathbf{r}_1|/\hbar}}{|\mathbf{r}-\mathbf{r}_1|}, \tag{3.7}$$

where g_π is a measure of the meson source strength of the nucleon. The gradient operator $\boldsymbol{\nabla}_1$ acts only on $\mathbf{r}_1$, so that (3.7) is evidently a solution of (3.6) (cf. (2.11)).

The 'dipole-like' nature of the field (3.7) is well understood by particle physicists, and the interaction energy between two nucleons associated with it is of 'dipole–dipole' form:

$$U_{12} \propto g_\pi^2(\boldsymbol{\sigma}_2 \cdot \nabla_2)(\boldsymbol{\sigma}_1 \cdot \nabla_1)\frac{e^{-mc|\mathbf{r}_2-\mathbf{r}_1|/\hbar}}{|\mathbf{r}_2-\mathbf{r}_1|}. \tag{3.8}$$

The π mesons are the mesons of smallest mass and hence give the largest contribution to the interaction at large distances. The appropriate length scale, from the exponential in (3.7), is

$$\hbar/mc \approx 1.4 \text{ fm}.$$

Explicit differentiation shows that (3.8) includes a potential of the tensor form $V_T(\mathbf{r})\Omega_T$. It is empirically established that π meson exchange is responsible for most of the tensor potential of (3.4), and is the dominant contribution to the whole potential at distances $|\mathbf{r}_2 - \mathbf{r}_1| > 1.4$ fm. At smaller distances other meson exchange processes become important, including the exchange of ρ mesons. However, the potentials at distances < 0.8 fm and, in particular, the short-range repulsion, are empirical and so far have no established explanation.

3.5 The weak interaction; β-decay

Hadrons are subject to the weak interaction as well as to the electromagnetic and strong interactions, and it is through the weak interaction that quarks, like leptons, are coupled to the W and Z bosons. For example, one quark can change to another by emitting or absorbing a virtual W boson. The phenomena of β-decay, in which a neutron becomes a proton or a proton becomes a neutron, proceed in this way.

In free space, the proton is the only stable three-quark system. The neutron in free space has enough excess mass over the proton to decay to it by the process shown in Fig. 3.3.

The mean life of the neutron in free space is 15.0 minutes. However, a neutron bound in a nucleus will be stable if the nuclear binding energies make decay energetically forbidden. Conversely, a proton bound in a nucleus may change into a neutron

$$p \rightarrow n + e^+ + \nu_e,$$

if the nuclear binding energies involved allow the process to occur. The

3.3 The decay $n \rightarrow p + e^- + \bar{\nu}_e$. As with muon decay, parity is not conserved in this weak interaction.

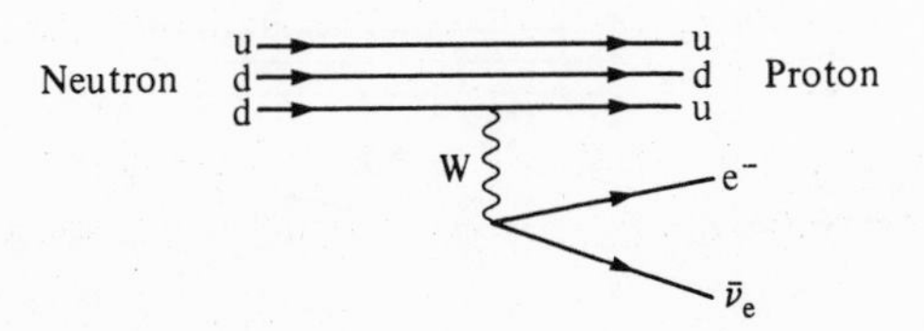

energetics of β-decay will be dealt with in detail in Chapter 4, and a more quantitative theory of β-decay will be given in Chapter 12.

3.6 More quarks

The u and d quarks are merely the two least massive of a sequence of types, or 'flavours' of quark, and to set the discussion of β-decay above into this wider context we list in Table 3.1 all the presently known flavours.

The existence of the more massive quarks in this table is revealed by the observation of states similar to the nucleon states and meson states we have already discussed, but which are apparently formed by substituting any of the 'new' quarks for the u or d quarks. Thus, for example, substituting an s quark for a d quark, there exists a K^+ meson ($u\bar{s}$) (mass 493.67 MeV/c^2) like the π^+ meson ($u\bar{d}$) but heavier, and a Σ^0 baryon (uds) (mass 1193 MeV/c^2) like the neutron (udd) but heavier. *Baryon* and *anti-baryon* are the generic names for particles essentially made up of three quarks or three anti-quarks. Again, since no quark has ever been isolated, the masses given in Table 3.1 are effective masses and have no precise significance.

Were it not for the weak interaction a heavy quark would be stable and there would be more absolute conservation laws, for example, the conservation of strangeness and the conservation of charm. Such laws hold for processes involving only the electromagnetic and strong interactions,

Table 3.1. *Properties of quarks*

Quark	Approximate mass (GeV/c^2)	Electric charge (e)
Down d	small	$-\frac{1}{3}$
Up u	small	$\frac{2}{3}$
Strange s	0.3	$-\frac{1}{3}$
Charm c	1.5	$\frac{2}{3}$
Bottom b	5.2	$-\frac{1}{3}$
Top t	~40.0?	$\frac{2}{3}$

3.4 The decays $\Sigma^- \rightarrow n + \pi^-$, $\Sigma^- \rightarrow n + \mu^- + \bar{\nu}_\mu$.

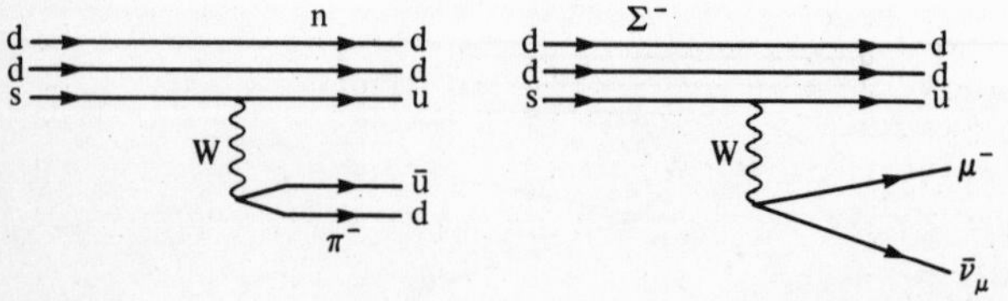

but are not absolute since all quarks couple to the $W^{\pm}$ and Z weak interaction fields, and a quark changes its flavour (but remains a quark!) when it emits or absorbs a virtual $W^{\pm}$ boson. Thus, for example, the s quark in the Σ^- baryon can decay through processes like those shown in Fig. 3.4. We shall see that nuclear binding energies are not sufficiently large to make a baryon containing a heavy quark stable even in a nucleus.

The weak interaction makes *all* mesons unstable. Mesons containing a heavy quark can decay by the heavy quark changing into a lighter quark. Another possible process is illustrated in Fig. 3.5, in which a quark and an anti-quark annihilate through the weak interaction into an anti-muon and a muon neutrino. This latter process is the predominant type of decay of the charged pions. The mean life of charged pions is 2.60×10^{-8} s.

The π^0 usually decays into two photons by the direct annihilation of the quarks with their own anti-quarks, in a way rather similar to the decay of positronium (an electron–positron pair e^+e^- in a bound state). Such a decay (Fig. 3.6) takes place through the electromagnetic interaction, and is therefore much quicker: the mean life of the π^0 is 0.83×10^{-16} s.

A baryon and an anti-baryon are always created or destroyed together. All the available experimental evidence is consistent with there being a law of '*conservation of baryon number*': the total number of baryons (anti-baryons being counted negatively) is conserved in all interactions.

3.5 The decay $\pi^+ \to \mu^+ + \nu_\mu$. The charged pion was discovered by Powell and co-workers in Bristol in 1947 by the observation of this decay.

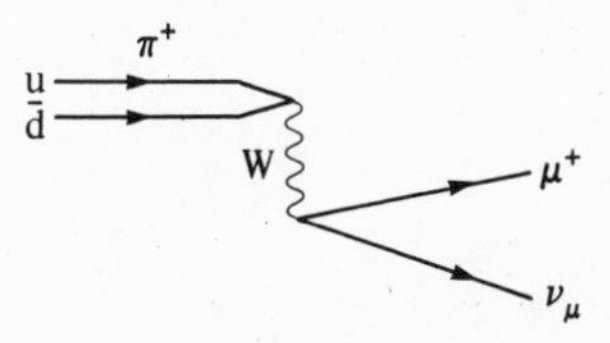

3.6 The electromagnetic decay. $\pi^0 \to \gamma + \gamma$.

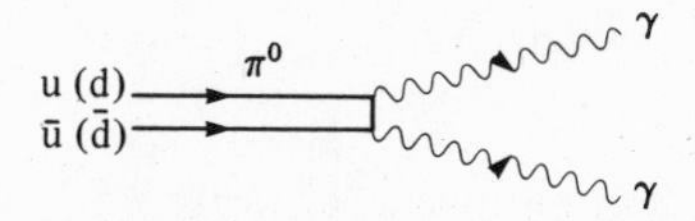

Problems

3.1 The spins of the neutron and the proton in the deuteron are aligned. Show that the magnetic moment of the deuteron is within 3% of the sum of the neutron and proton moments. What might be the origin of the discrepancy?

3.2(*a*) Show that the magnetic interaction energy between two magnetic dipoles $\mu\boldsymbol{\sigma}_1$ and $\mu\boldsymbol{\sigma}_2$ is of the form $V_T(r)\Omega_T$ with $V_T(r) = -(\mu_0/4\pi)\mu^2/r^3$. ($\mu_0$ is the permeability of the vacuum.)

(*b*) Verify that equation (3.8) includes terms in the nucleon–nucleon potential of tensor form.

3.3 The Coulomb self-energy of a hadron with charge +e or −e is about 1 MeV. The quark content and rest energies (in MeV) of some hadrons are:

n(udd) 940, p(uud) 938

Σ^-(dds) 1197, Σ^0(uds) 1192, Σ^+(uus) 1189

K^0(d$\bar{s}$) 498, K^+(u$\bar{s}$) 494.

The u and d quarks make different contributions to the rest energy. Estimate this difference.

3.4 Which of the following processes are allowed by the conservation laws:

(*a*) $n \to p + \gamma$,

(*b*) $p \to e^+ + \gamma$,

(*c*) $p \to \pi^+ + \gamma$,

(*d*) $\bar{p} + n \to \pi^- + \pi^0$?

3.5 The decay Ξ^{*-} initiates the sequence of decays shown below:

$\Xi^{*-} \to K^- + \Sigma^0$

$\quad \Sigma^0 \to \Lambda^0 + \gamma$

$\quad\quad \Lambda^0 \to p + e^- + \bar{\nu}_e$

$\quad K^- \to \pi^- + \pi^0$

$\quad\quad \pi^0 \to \gamma + \gamma$

$\quad\quad \pi^- \to \mu^- + \bar{\nu}_\mu$

$\quad\quad\quad \mu^- \to e^- + \bar{\nu}_e + \nu_\mu$

The quark content of the hadrons involved is:

Ξ^{*-}(ssd), Σ^0(sud), Λ^0(sud), p(uud),

K^-(s, $\bar{u}$), π^-(d$\bar{u}$), π^0(u$\bar{u}$ − d$\bar{d}$).

Classify the decays as strong, electromagnetic, or weak.

4

Nuclear sizes and nuclear masses

We now turn to the study of the nucleus. A nucleus is a bound assembly of neutrons and protons. ${}^{A}_{Z}X$ denotes a nucleus of an atom of the chemical element X containing A nucleons, of which Z are protons and $N = (A - Z)$ are neutrons. For example, ${}^{35}_{17}Cl$ denotes a chlorine nucleus with 18 neutrons and ${}^{37}_{17}Cl$ a chlorine nucleus with 20 neutrons. Since the chemical symbol determines the atomic number Z, ${}^{35}Cl$ or ${}^{37}Cl$ is identification enough, but the addition of the Z label is often useful. $A = (N + Z)$ is called the *mass number* of the nucleus. Nuclei which differ only in the number of neutrons they contain are called *isotopes*. Nuclei of the same A but different Z are called *isobars*.

4.1 Electron scattering by the nuclear charge distribution

Rutherford's famous analysis in 1911 of the scattering of α-particles by matter established that the size of the nucleus of an atom is small compared with the size of the atom. Whereas the electronic distribution extends to a distance of the order of angströms ($1\ \text{Å} = 10^{-10}$ m) from the nucleus, these and later experiments showed that the distribution of nucleons is confined to a few fermis ($1\ \text{fm} = 10^{-15}$ m). Early theories of α-decay and nuclear binding energies gave estimated values for nuclear radii of a similar magnitude.

Precise information came in the 1950s, with experiments using the elastic scattering of high-energy electrons to probe the nuclear charge distribution.

There is an obvious advantage in using charged leptons (electrons or muons) to probe nuclear matter, since leptons interact with nucleons primarily through electromagnetic forces: the complications of the strong nuclear interaction are not present, and the weak interaction is negligible for the scattering process. The most significant interaction between a charged lepton, which can be regarded as a structureless point object, and the nuclear charge, is the Coulomb force, and this is well understood. If the nucleus has a magnetic moment, the magnetic contribution to the scattering becomes important at large scattering angles, but this also is well understood.

If scattering experiments are to give detailed information on the nuclear charge distribution, it is clear that the de Broglie wavelength λ of the incident particle must be less than, or at least comparable with, the distances over which the nuclear charge density changes. An electron with $(\lambda/2\pi) \sim 1$ fm has momentum $p = 2\pi\hbar/\lambda$ and hence energy $E = (p^2c^2 + m^2c^4)^{\frac{1}{2}} \sim 200$ MeV. At these energies, the electrons are described by the Dirac relativistic wave equation, rather than by the Schrödinger equation. The experiments yield a differential cross-section $\mathrm{d}\sigma(E, \theta)/\mathrm{d}\Omega$ (Appendix A) for elastic scattering from the nucleus through an angle θ, which depends on the energy E of the incident electrons. Typical experimental data are shown in Fig. 4.1.

The incident electrons are, of course, also scattered by the atomic electrons in the target. However, this scattering is easily distinguished from the nuclear scattering by the lower energy of the scattered electrons. Whereas the recoil energy taken up by the heavy nucleus is very small, the recoil energy taken up by the atomic electrons is appreciable, except for scattering in the forward direction. (See Problem 4.1.)

The nuclear charge density will be described by some density function $e\rho_{\mathrm{ch}}(\mathbf{r})$. (The proton charge e is put in as a factor for convenience.) This function is not necessarily spherically symmetric – we shall mention this later – but for nuclei which are spherically symmetric, or nearly so, we can assume the charge density depends only on the distance r from the centre of the nucleus. Then, using the Dirac wave equation for the electron, $\mathrm{d}\sigma/\mathrm{d}\Omega$ is in principle completely determined by $\rho_{\mathrm{ch}}(r)$, though the calculations are not trivial. The inverse problem, that of finding $\rho_{\mathrm{ch}}(r)$ from a knowledge of $\mathrm{d}\sigma/\mathrm{d}\Omega$, is even more difficult (see Problem 4.2). The restricted amount of experimental information available means that, at best, only a partial resolution of the problem can be made. Some idea of the results of a direct inversion of scattering data is given by Fig. 4.2.

It has been more usual to assume a plausible shape for $\rho_{\mathrm{ch}}(r)$, describe this

4.1 Experimental elastic electron-scattering differential cross-section from gold $^{197}_{79}Au$ at energies of 126 MeV and 183 MeV. The fitted curves are calculated with an assumed charge distribution of the form given by equation (4.1), with $R=6.63$ fm, $a=0.45$ fm. The cross-section to be expected, at 126 MeV, if the gold nucleus had a point charge is shown for comparison. (Data and theoretical curves taken from Hofstadter, R. (1963), *Electron Scattering and Nuclear and Nucleon Structure*, New York: Benjamin.)

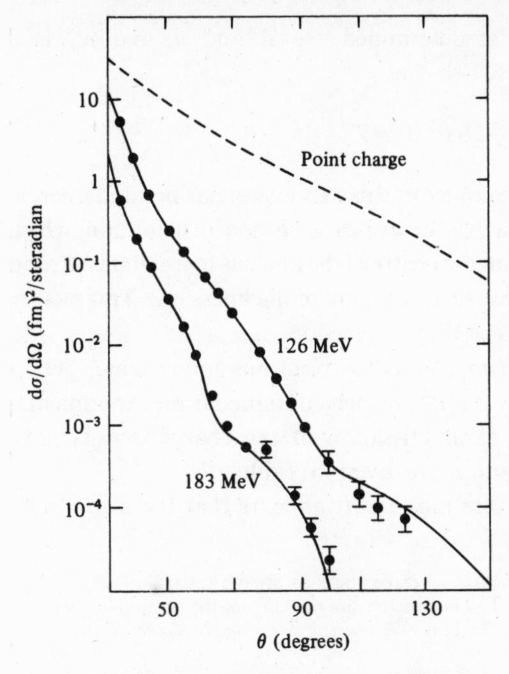

4.2 The electric charge density of $^{208}_{82}Pb$ from a model-independent analysis of electron scattering data. The bars indicate the uncertainty. (Friar, J. L. & Negele, J. W. (1973), *Nuc. Phys.* **A212**, 93.)

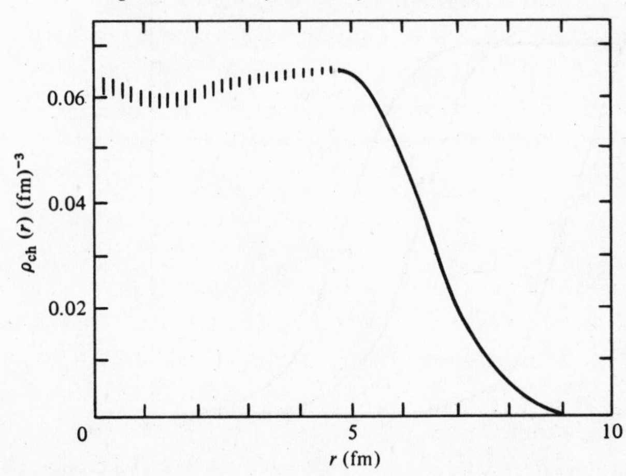

by a simple mathematical expression involving a few parameters, and then determine the parameters by fitting to the scattering data. A form which has been widely adopted is

$$\rho_{\text{ch}}(r) = \frac{\rho^0_{\text{ch}}}{1 + e^{(r-R)/a}}, \tag{4.1}$$

where the parameters to be determined are R and a, and ρ^0_{ch} is a normalisation constant chosen so that

$$\int \rho_{\text{ch}}(r)\, d^3\mathbf{r} = 4\pi \int_0^\infty \rho_{\text{ch}}(r) r^2\, dr = Z.$$

It should be stressed that the choice of this expression has no fundamental significance; it just conveniently describes a charge distribution which extends almost uniformly from the centre of the nucleus to a distance R, and falls to zero over a well-defined surface region of thickness $\sim a$. This picture is consistent with the results of direct inversion.

In Fig. 4.3 we show nuclear charge distributions for a light ($^{16}_{8}$O), a medium ($^{109}_{47}$Ag) and a heavy ($^{208}_{82}$Pb) nucleus obtained from experimental scattering data, using this parametrisation of the charge density. The corresponding values of R and a are given in Table 4.1.

As the examples in the table indicate, it appears that there is a well-

4.3 The electric charge density of three nuclei as fitted by $\rho_{\text{ch}}(r) = \rho^0_{\text{ch}}/[1 + \exp((r-R)/a)]$. The parameters are taken from the compilation in Barrett, R. C. & Jackson, D. F. (1977), *Nuclear Sizes and Structure*, Oxford: Clarendon Press.

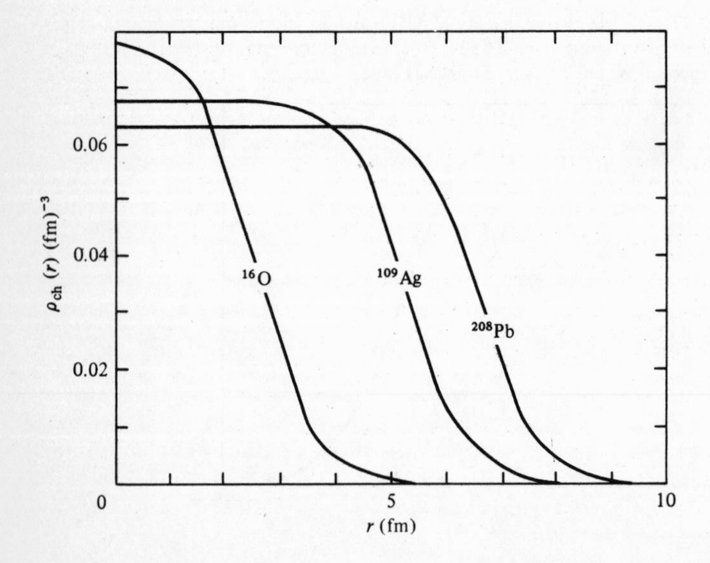

defined 'surface region' which has much the same width for all nuclei, even light ones.

4.2 Muon interactions

The negative muon is another leptonic probe of nuclear charge. Its properties, other than its mass of $m_\mu \approx 207\, m_e$ and its mean life of 2.2×10^{-6} s, are similar to those of the electron. However, the radius of its lowest Bohr orbit in an atom of charge Z is $(4\pi\varepsilon_0)\hbar^2/m_\mu Ze^2$, and this is smaller than the corresponding electron orbit by a factor (m_e/m_μ). For $Z=50$ the radius is only 5 fm. Hence the wave-functions of the lowest muonic states will lie to a considerable extent within the distribution of nuclear charge, particularly in heavy nuclei, and the energies of these states will therefore depend on the details of the nuclear charge distribution.

Experimentally, negative muons are produced in the target material by the decay of a beam of negative pions, and are eventually captured in outer atomic orbitals. Before they decay, many muons fall into lower orbits, emitting X-rays in the transitions. The measured energies of these X-rays may be compared with those calculated with various choices of parameters for $\rho_{ch}(r)$. Values of R and a, found in this way, agree well with results from electron scattering.

4.3 The distribution of nuclear matter in nuclei

From the distribution of charge in a nucleus, which as we have seen can be determined by experiment, we can form some idea of the distribution of nuclear matter. If the proton were a point object, we could identify the proton number density $\rho_p(r)$ with $\rho_{ch}(r)$. Since the strong nuclear forces which bind nucleons together are charge independent and of short range, we can assume that to a good approximation the ratio of neutron density ρ_n to proton density ρ_p is the same at all points in a nucleus, i.e. $\rho_n(r)/\rho_p(r) = N/Z$. Then the total density of nucleons $\rho = \rho_n + \rho_p$ can be expressed as $\rho =$

Table 4.1. *Nuclear radii (R) and nuclear surface widths (a)*

Nucleus	R (fm)	a (fm)	$R/A^{\frac{1}{3}}$ (fm)
$^{16}_{8}$O	2.61	0.513	1.04
$^{109}_{47}$Ag	5.33	0.523	1.12
$^{208}_{82}$Pb	6.65	0.526	1.12

$(A/Z)\rho_{ch}$, where $A=N+Z$. The resulting nuclear matter densities for the same nuclei we took in Fig. 4.3 are plotted in Fig. 4.4. These densities are only approximate, since we have neglected the finite size of both proton and neutron and the effect of Coulomb forces, but they indicate that at the centre of a nucleus the nuclear-matter density ρ is roughly the same for all nuclei. It increases with A, but appears to tend to a limiting value ρ_0 of about 0.17 nucleons fm^{-3} for large A. The existence of this limiting value ρ_0, known as the 'density of nuclear matter', is an important result. Consistently with this, we find (Table 4.1), that the 'radius' R of a nucleus is very closely proportional to $A^{\frac{1}{3}}$, and, approximately, $(4\pi/3)R^3\rho_0=A$. We shall take

$$\rho_0=0.17 \text{ nucleons fm}^{-3}$$

and (4.2)

$$R=1.1A^{\frac{1}{3}}\text{ fm}.$$

4.4 The masses and binding energies of nuclei in their ground states

It thus appears that a nucleus is rather like a spherical drop of liquid, of nearly uniform density. How are we to understand its properties? A nucleus is a quantum-mechanical system. We shall see later that its excited states are generally separated by energies ~ 1 MeV from its ground state, so that to all intents and purposes nuclei in matter at temperatures

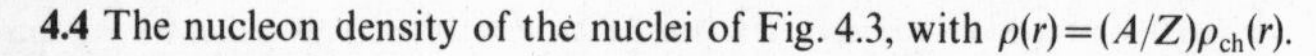

4.4 The nucleon density of the nuclei of Fig. 4.3, with $\rho(r)=(A/Z)\rho_{ch}(r)$.

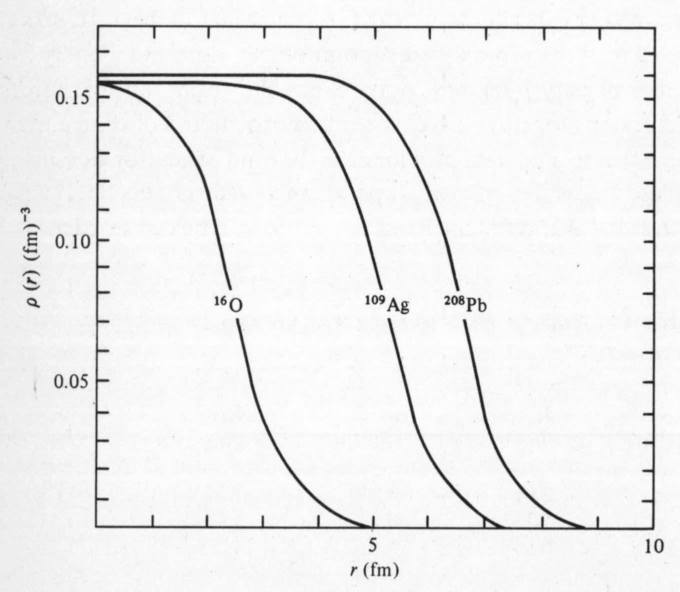

that are accessible on Earth are in their ground states. Like any other finite system, a nucleus in its ground state has a well-defined energy and a well-defined angular momentum. In this chapter we shall be concerned with the ground-state energy. Other ground-state properties of a nucleus will be discussed in the next chapter.

Since a nucleus is a bound system, an energy $B(Z, N)$ is needed to pull it completely apart into its Z protons and N neutrons. From the Einstein relation between mass and energy, the *binding energy* $B(Z, N)$ is related to the mass $m_{\text{nuc}}(Z, N)$ of the nucleus by

$$m_{\text{nuc}}(Z, N) = Zm_{\text{p}} + Nm_{\text{n}} - B(Z, N)/c^2, \tag{4.3}$$

and $B(Z, N)$ must be positive for the nucleus to be formed. We shall see that nuclear binding energies are of the order of 1% of the rest-mass energy $m_{\text{nuc}}c^2$.

Experimentally, the masses of atomic ions, rather than the masses of bare nuclei, are the quantities usually measured directly. If $m_a(Z, N)$ is the mass of the neutral atom,

$$m_a(Z, N) = Z(m_{\text{p}} + m_{\text{e}}) + Nm_{\text{n}} - B(Z, N)/c^2 - b_{\text{electronic}}/c^2, \tag{4.4}$$

where $b_{\text{electronic}}$ is the binding energy of the atomic electrons. These electronic contributions are, for many purposes, negligible. (The simple Thomas–Fermi statistical model of a neutral atom gives the total electronic binding energy $\approx 20.8Z^{\frac{7}{3}}$ eV.)

Atomic masses are known very accurately, and published tables give atomic masses rather than nuclear masses. Measurements in 'mass spectrometers' depend on the deflection of charged ions in electric and magnetic fields. Instruments of great ingenuity have been developed, giving relative masses accurate to about one part in 10^7. The unit employed is the *atomic mass unit*, which is defined to be $\frac{1}{12}$ of the mass of the neutral ^{12}C atom:

$$1\ \text{amu} = 931.5016 \pm 0.0026\ \text{MeV}/c^2.$$

Differences between the masses of stable atoms and unstable, radioactive, atoms (for which mass spectrometers may be inappropriate) can be determined by measuring the energy release in the unstable atom decay, again using the Einstein mass–energy relation.

Table 4.2 shows the experimental binding energies for some of the lighter nuclei, those formed by successively adding a proton followed by a neutron to an original neutron. Note that all the binding energies are positive: this reflects the basic long-range attraction of the nucleon–nucleon interaction.

Also given in the table is the average binding energy per nucleon,

$B(Z, N)/A$. For the heavier nuclei in the table, the average binding energy appears to be gradually increasing to around 8 MeV, but the numbers fluctuate somewhat from nucleus to nucleus. The fluctuation is more dramatically exhibited in the binding energy difference between a nucleus and the one preceding it, also shown in the table. This energy can be interpreted as the binding energy of the last nucleon added to the nucleus in the given sequence. It is particularly large for the 'even–even' nuclei $^4_2$He, $^8_4$Be, $^{12}_6$C and $^{16}_8$O, and particularly small for the nuclei immediately following, growing steadily as the next three nucleons are added to form the next even–even nucleus. Clearly we see here some extra binding energy associated with neutron–neutron and proton–proton pairing. The effect stems from the attractive character of the nucleon–nucleon interaction, and is associated with the pairing of angular momenta discussed in Chapter 5. Table 4.2 also gives the spins and parities of the nuclei for later reference; it will be seen that the even–even nuclei have spin zero.

As we shall see in Chapter 6, because of its low mass, low electric charge, and relatively large binding energy, the first even–even nucleus $^4_2$He is particularly important in the nuclear physics of heavy nuclei. Indeed, $^4_2$He

Table 4.2. *Energies of some light nuclei (MeV)*

Nucleus	Binding energy (MeV)	Binding energy of last nucleon (MeV)	Binding energy per nucleon (MeV)	Spin and parity
$^2_1$H	2.22	2.2	1.1	1^+
$^3_1$H	8.48	6.3	2.8	$\frac{1}{2}^+$
$^4_2$He	28.30	19.8	7.1	0^+
$^5_2$He	27.34	−1.0	5.5	$\frac{3}{2}^-$
$^6_3$Li	31.99	4.7	5.3	1^+
$^7_3$Li	39.25	7.3	5.6	$\frac{3}{2}^-$
$^8_4$Be	56.50	17.3	7.1	0^+
$^9_4$Be	58.16	1.7	6.5	$\frac{3}{2}^-$
$^{10}_5$B	64.75	6.6	6.5	3^+
$^{11}_5$B	76.21	11.5	6.9	$\frac{3}{2}^-$
$^{12}_6$C	92.16	16.0	7.7	0^+
$^{13}_6$C	97.11	5.0	7.5	$\frac{1}{2}^-$
$^{14}_7$N	104.66	7.6	7.5	1^+
$^{15}_7$N	115.49	10.8	7.7	$\frac{1}{2}^-$
$^{16}_8$O	127.62	12.1	8.0	0^+
$^{17}_8$O	131.76	4.1	7.8	$\frac{5}{2}^+$

played an important role in the early history of nuclear physics and before it was properly identified it was given a special name, the α-particle, a name still in use today.

Some of the large binding energy of the nuclei ${}^4_2$He, ${}^{12}_6$C and ${}^{16}_8$O can be associated with their 'shell structure', which will be discussed in Chapter 5. As for ${}^8_4$Be, its binding energy is less than that of two α-particles by 0.1 MeV, and so the nucleus ${}^8_4$Be is unstable. It does have a transient existence for a time long compared with the 'nuclear time-scale' (§ 5.2), but if it is formed it will eventually fall apart into two α-particles.

Another interesting special case in Table 4.2 is that of ${}^5_2$He. The binding energy of the last nucleon is here negative; if ${}^5_2$He is formed it, too, has only a transient existence before falling apart into a neutron and an α-particle. The other nuclei in Table 4.2 are all stable.

4.5 The semi-empirical mass formula

The features of 'pairing energies' and shell-structure effects, superposed on a slowly varying binding energy per nucleon, can be discerned throughout the range of nuclei for which data are available. We saw in §4.3 that the density of nuclear matter is approximately constant, and also that nuclei have a well-defined surface region. It appears as if a nucleus behaves in some ways rather like a drop of liquid. This analogy is made more precise in the '*semi-empirical mass formula*', a remarkable formula which, with just a few parameters, fits the binding energies of all but the lightest nuclei to a high degree of accuracy. There are several versions of the mass formula. The one which is sufficiently accurate for the purposes of this book gives for the total binding energy of a nucleus of A nucleons, made up of Z protons and N neutrons,

$$B(N,Z)=aA-bA^{\frac{2}{3}}-s\frac{(N-Z)^2}{A}-\frac{dZ^2}{A^{\frac{1}{3}}}-\frac{\delta}{A^{\frac{1}{2}}}. \tag{4.5}$$

The parameters a, b, s, d and δ are found by fitting the formula to measured binding energies. Wapstra (*Handbuch der Physik*, XXXVIII/1) gives

$$a=15.835 \text{ MeV}$$

$$b=18.33 \text{ MeV}$$

$$s=23.20 \text{ MeV}$$

$$d=0.714 \text{ MeV}$$

and

$$\delta=\begin{cases} +11.2 \text{ MeV for odd–odd nuclei (i.e., odd } N\text{, odd } Z) \\ 0 \text{ for even–odd nuclei (even } N\text{, odd } Z\text{, or even } Z\text{, odd } N) \\ -11.2 \text{ MeV for even–even nuclei (even } N\text{, even } Z). \end{cases}$$

It is the first two terms in this formula which have an analogue in the theory of liquids. The term (aA) represents a constant bulk-binding energy per nucleon, like the cohesive energy of a simple liquid. The second term represents a surface energy, in particular the surface energy of a sphere. The surface area of a sphere is proportional to the two-thirds power of its volume and hence, at constant density of nucleons, to $A^{\frac{2}{3}}$. As in a liquid, this term subtracts from the bulk binding since the particles in the surface are not in the completely enclosed attractive environment of those in the bulk. In liquids this term is identified with the energy of surface tension, and is responsible for drops of liquid being approximately spherical when gravitational effects are small. In nuclei, gravitational effects are always small, and indeed nuclei do tend to be spherical.

The term $-\mathrm{d}Z^2/A^{\frac{1}{3}}$, called the Coulomb term, also has a simple explanation; it is the electrostatic energy of the nuclear charge distribution. If the nucleus were a uniformly charged sphere of radius $R_0A^{\frac{1}{3}}$ (equation (4.2)) and total charge Ze, it would have energy

$$E_{\mathrm{c}}=\frac{3}{5}\frac{(Ze)^2}{(4\pi\varepsilon_0)R_0A^{\frac{1}{3}}}. \tag{4.6}$$

With $R_0=1.1$ fm this gives an estimate of d, $d=0.79$ MeV, close to the value found empirically.

The term $-s(N-Z)^2/A$ is the simplest expression which, by itself, would give the maximum binding energy, for fixed A, when $N=Z$ (A even) or $N=Z\pm 1$ (A odd). It is called the symmetry energy, since it tends to make nuclei symmetric in the number of neutrons and protons. As was exemplified in the case of the deuteron discussed in Chapter 3, the average neutron–proton attraction in a nucleus is greater than the average neutron–neutron or proton–proton attraction, essentially as a consequence of the Pauli exclusion principle. Thus for a given A it is energetically advantageous to maximise the number of neutron–proton pairs which can interact: this is achieved by making Z and N as near equal as possible. Since the forces are short range, the term must correspond to a 'bulk' effect, like the cohesive energy. Hence there must be a factor A in the denominator, so that overall the term is proportional to A for a fixed ratio of neutrons to protons. One can also argue (see Problem 5.2) that the kinetic energy contribution to the energy results in a similar term, which is absorbed in the coefficient s.

The final term in the semi-empirical mass formula is the pairing energy $\delta/A^{\frac{1}{2}}$, manifest in the light nuclei included in Table 4.2. It is purely phenomenological in form and the $A^{-\frac{1}{2}}$ dependence is empirical. For the larger nuclei the pairing energy is small but, as we shall see, it does give rise to important physical effects.

More sophisticated versions of the formula include also 'shell structure' effects (Chapter 5), but for nuclei heavier than neon ($A=20$) for which our formula is appropriate these extra terms are of less significance than the five terms of equation (4.5).

We have in the semi-empirical mass formula a description and an understanding of the binding energies of the nuclei. We shall see that it gives a simple but profound explanation of the masses of the chemical elements and of why there is only a finite number of stable atoms in chemistry.

4.6 The β-stability valley

Using equations (4.3) and (4.5), the mass of the neutral atom with its nucleus having Z protons and N neutrons is given by

$$m_a(N,Z)c^2=(Nm_\mathrm{n}+Z(m_\mathrm{p}+m_\mathrm{e}))c^2-aA+bA^{\frac{2}{3}}+\frac{\mathrm{d}Z^2}{A^{\frac{1}{3}}}$$

$$+\frac{s(N-Z)^2}{A}+\frac{\delta}{A^{\frac{1}{2}}}, \tag{4.7}$$

(neglecting the electron binding energies).

For a fixed number of nucleons A, we can write this as a function of Z, replacing N by $A-Z$:

$$\begin{aligned} m_a(A,Z)c^2 &= (Am_\mathrm{n}c^2-aA+bA^{\frac{2}{3}}+sA+\delta A^{-\frac{1}{2}}) \\ &\quad -(4s+(m_\mathrm{n}-m_\mathrm{p}-m_\mathrm{e})c^2)Z+(4sA^{-1}+\mathrm{d}A^{-\frac{1}{3}})Z^2 \\ &= \alpha-\beta Z+\gamma Z^2, \quad \text{say.} \end{aligned} \tag{4.8}$$

Consider first the case A odd, so that $\delta=0$. The plot of $m_a(A,Z)$ against Z is a parabola, with a minimum at

$$Z=\beta/2\gamma=\frac{(4s+(m_\mathrm{n}-m_\mathrm{p}-m_\mathrm{e})c^2)A}{2(4s+\mathrm{d}A^{\frac{2}{3}})} \tag{4.9}$$

Thus the atom with the lowest rest-mass energy for given A has Z equal to the integer $Z_{\min}$ closest to $\beta/2\gamma$. From the form of the expression (4.9) and the values of the parameters, it is evident that $Z_{\min}\leqslant A/2$, so that $N\geqslant Z$ for this nucleus.

Now β-decay, described in § 3.5, is a process whereby the Z of a nucleus changes while A remains fixed, if the process is energetically allowed. Thus if a nucleus has $Z<Z_{\min}$ the process

$$(A,Z)\to(A,Z+1)+\mathrm{e}^-+\bar{\nu}_\mathrm{e}$$

is possible if

$$m_{\text{nuc}}(A, Z) > m_{\text{nuc}}(A, Z+1) + m_e, \tag{4.10}$$

since the mass of the anti-neutrino (if indeed it has mass) is exceedingly small. Adding Zm_e to each side of this inequality, the condition may be written in terms of atomic masses:

$$m_a(A, Z) > m_a(A, Z+1). \tag{4.11}$$

More precisely, conditions (4.10) and (4.11) differ by a few (electron volts)/c^2, associated with the electronic binding energy differences, and since β-decay usually takes place in an atomic environment (4.11) is the more suitable form. The energy released in nuclear β-decay is never large enough to produce particles other than electrons or positrons, and neutrinos.

As an example, $^{77}_{32}\text{Ge}$ decays by a series of β-decays to $^{77}_{34}\text{Se}$, Z increasing by one at each stage:

$$\begin{array}{l} ^{77}_{32}\text{Ge} \rightarrow {}^{77}_{33}\text{As} + \text{e} + \bar{\nu}_e + 2.75\ \text{MeV} \\ \qquad\qquad\quad \downarrow \\ \qquad\qquad {}^{77}_{34}\text{Se} + \text{e} + \bar{\nu}_e + 0.68\ \text{MeV}. \end{array}$$

$^{77}_{34}\text{Se}$ is the only stable nucleus with $A = 77$.

A nucleus with $Z > Z_{\text{min}}$ can decay by emitting a positron and a neutrino. For example, another sequence of decays ending in $^{77}_{34}\text{Se}$ is:

$$\begin{array}{l} ^{77}_{36}\text{Kr} \rightarrow {}^{77}_{35}\text{Br} + \text{e}^+ + \nu_e + 2.89\ \text{MeV} \\ \qquad\qquad\quad \downarrow \\ \qquad\qquad {}^{77}_{34}\text{Se} + \text{e}^+ + \nu_e + 1.36\ \text{MeV}. \end{array}$$

For the process of β-decay by positron emission to be possible the condition is

$$m_{\text{nuc}}(A, Z) > m_{\text{nuc}}(A, Z-1) + m_e,$$

or, in terms of atomic masses,

$$m_a(A, Z) > m_a(A, Z-1) + 2m_e. \tag{4.12}$$

In an atomic environment, a β-decay process competing with positron emission is *electron capture*, in which the nucleus absorbs one of its cloud of atomic electrons, emitting only a neutrino. For example,

$$^{77}_{35}\text{Br} + \text{e} \rightarrow {}^{77}_{34}\text{Se} + \nu_e + 2.38\ \text{MeV}.$$

Such processes are often referred to as K-capture, since the electron is most likely to come from the innermost 'K-shell' of atomic electrons. The condition for K-capture to be possible is less restrictive than (4.12):

$$m_{\text{nuc}}(A, Z) + m_e > m_{\text{nuc}}(A, Z-1),$$

or

$$m_a(A, Z) > m_a(A, Z-1). \tag{4.13}$$

For example, ${}^{7}_{4}\mathrm{Be}$ decays by K-capture:

$${}^{7}_{4}\mathrm{Be} + \mathrm{e} \rightarrow {}^{7}_{3}\mathrm{Li} + \nu_{\mathrm{e}} + 0.86\ \mathrm{MeV},$$

whereas it cannot decay by positron emission. When both processes are possible, the energy release in K-capture will be $2m_{\mathrm{e}}c^2 \approx 1$ MeV greater than in the corresponding positron emission.

Thus odd-A nuclei decay to the value of Z closest to $\beta/2\gamma$. It is clearly highly unlikely that there will be two values of Z giving exactly the same atomic masses; we expect there to be only one β-stable Z value for odd-A nuclei, and such is the case.

Nuclei with even A must have Z and N both even numbers, or Z and N both odd numbers. In the semi-empirical mass formula, the even–even nuclei have a lower energy than the odd–odd nuclei by $2\delta A^{-\frac{1}{2}}$. This quantity varies from 5 MeV when $A = 20$ to 1.4 MeV when $A = 250$. Thus there are two mass parabolas with relative vertical displacement $2\delta A^{-\frac{1}{2}}/c^2$, as in Fig. 4.5, for each even value of A.

In Fig. 4.5, the values $Z = 28$ and $Z = 30$ on the lower even–even parabola are both stable with respect to β-decay, since processes in which two electrons or two positrons are emitted simultaneously have not been observed. The figure is characteristic of nuclei with even A, and pairs of stable nuclei with different (even) Z but the same A are common. The only odd–odd nuclei which are stable are the four lightest: ${}^{2}_{1}\mathrm{H}$, ${}^{6}_{3}\mathrm{Li}$, ${}^{10}_{5}\mathrm{B}$ and ${}^{14}_{7}\mathrm{N}$ – but for $A < 20$ the semi-empirical mass formula is less accurate.

The nuclei which are observed to be β-stable are plotted in Fig. 4.6 as points in the (N, Z) plane. Nuclei of constant A lie on the diagonal lines $N + Z = A$. The bottom of the 'β-stability valley' where the β-stable nuclei are found is given remarkably well by the approximation (equation (4.9)).

$$Z = \beta/2\gamma = \frac{(4s + (m_{\mathrm{n}} - m_{\mathrm{p}} - m_{\mathrm{e}})c^2)A}{2(4s + \mathrm{d}A^{\frac{2}{3}})}. \tag{4.14}$$

4.7 The masses of the β-stable nuclei

With the approximation $Z = \beta/2\gamma$, the binding energies of the β-stable nuclei can be calculated from equation (4.5). Neglecting the pairing energy, the resulting binding energy per nucleon $B(A)/A$ is plotted against A in Fig. 4.7 and the various contributions to $B(A)/A$ are displayed in Fig. 4.8.

It should be noted that apart from pairing effects the bulk term is the only positive contribution to the binding energy. The initial rise of B/A with A is simply due to the negative surface contribution diminishing in magnitude relative to the bulk contribution as the size of the nucleus increases. However, as A and therefore Z increase further, the quadratic Coulomb term becomes important and produces a maximum on the curve.

The curve gives the observed nuclear-binding energies quite well. The small deviations of the experimental values from the smooth curve are for the most part due to the quantum mechanical 'shell' effects, which are considered in the next chapter. The maximum binding energies lie in the neighbourhood of ^{56}Fe.

4.8 The energetics of α-decay and fission

The peak in the binding energy curve makes possible other modes of decay for a heavy nucleus which is stable against β-decay. Since there is a gradual decrease of (B/A) with A for the heavier nuclei, it may be energetically advantageous for a heavy nucleus to split into two smaller nuclei, which together have a greater net binding energy. The most common

4.5 The atomic masses of atoms with $A = 64$ relative to the atomic mass of $^{64}_{28}$Ni. Open circles ○ are odd–odd nuclei, filled circles ● are even–even nuclei. The theoretical even–even and odd–odd parabolas are drawn using the parameters of equation (4.5). Note the odd–odd nucleus $^{64}_{29}$Cu, which can β^--decay to $^{64}_{30}$Zn or β^+-decay to $^{64}_{28}$Ni, both of which are stable, naturally occurring, isotopes. These decays are discussed in detail in Chapter 12.

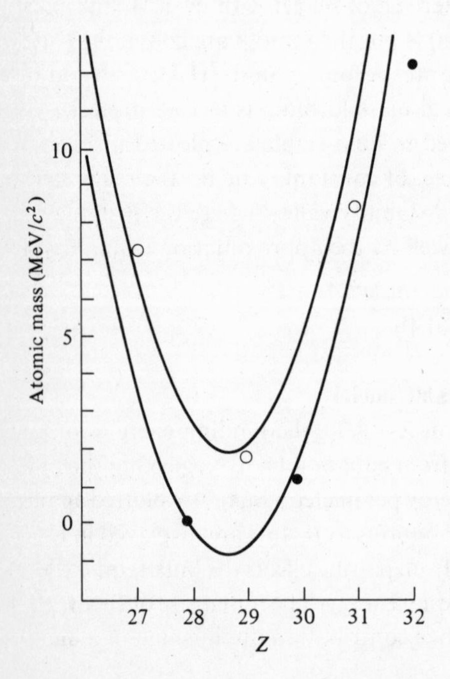

such process is the emission of an α-particle. As Table 4.2 shows, $^4_2$He has the comparatively large binding energy of 28.3 MeV. The condition for α-emission to be possible from a nucleus (A, Z) to give a nucleus $(A-4, Z-2)$

4.6 The β-stability valley. Filled squares denote the stable nuclei and long-lived nuclei occurring in nature. Neighbouring nuclei are unstable. Those for which data on masses and mean lives are known fill the area bounded by the lines. For the most part these unstable nuclei have been made artificially. (Data taken from *Chart of the Nuclides* (1977), Schenectady: General Electric Company.)

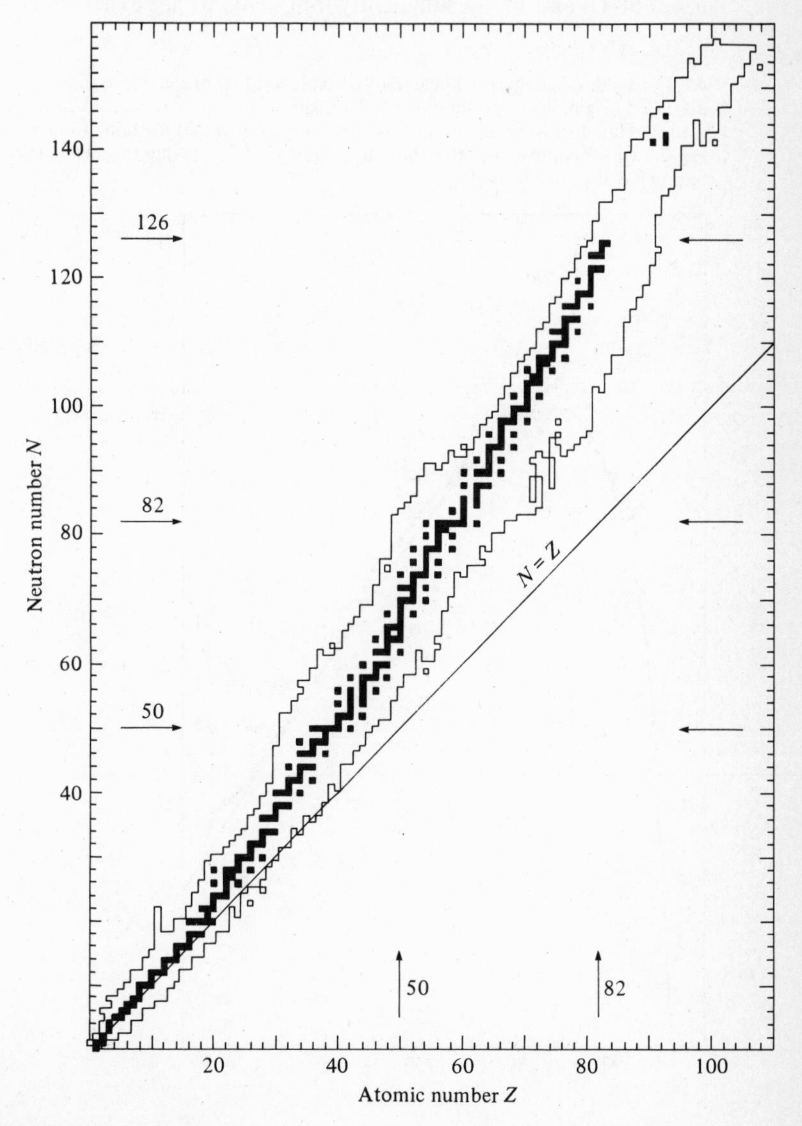

is

$$B(A,Z) < B(A-4, Z-2) + 28.3 \text{ MeV}. \tag{4.15}$$

For (A, Z) on the line of β-stability, this condition is always satisfied for sufficiently large A, $A \gtrsim 165$, and all such nuclei are, in principle, able to emit α-particles. However, we shall see in Chapter 6, where the physical mechanism of α-decay is analysed, that decay rates are so slow that the β-stable nuclei can also be regarded as α-stable up to $^{209}_{83}$Bi. Beyond, only some isotopes of Th and U are sufficiently long-lived to have survived on

4.7 The binding energy per nucleon of β-stable (odd-A) nuclei. Note the displaced origin. The smooth curve is from the semi-empirical mass formula with Z related to A by equation (4.14). Experimental values for odd-A nuclei are shown for comparison; the main deviations ($<1\%$) are due to 'shell' effects not included in our formula.

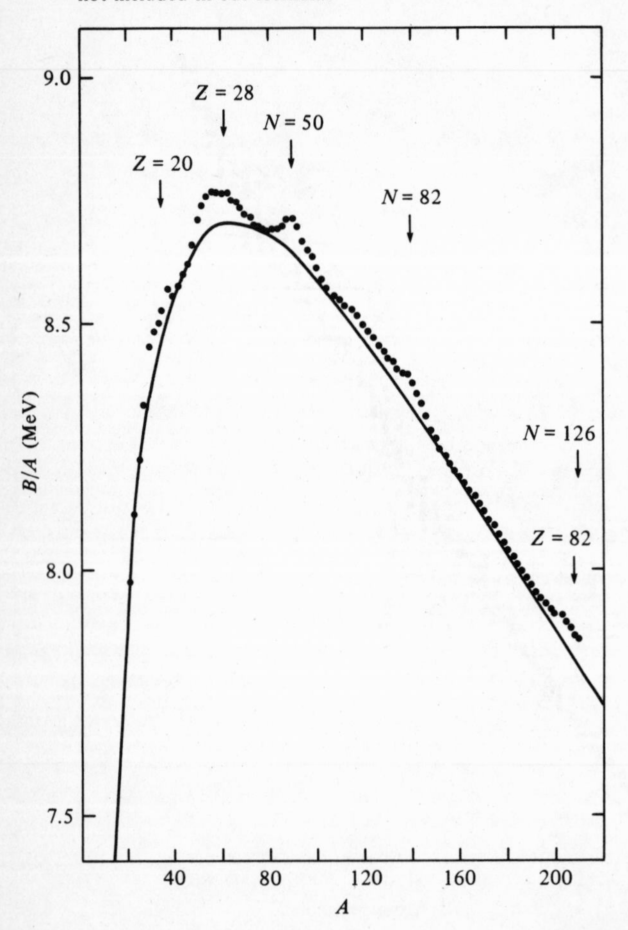

Earth since its formation; other unstable heavy elements are produced either from the decay of these, or artificially.

Another energetically-favourable process which is possible when A is large is the splitting of a nucleus into two more nearly equal parts. This is called *fission*. The energetics of fission may be explored using the semi-empirical mass formula, and in Chapter 6 we shall investigate the rate of spontaneous fission processes.

Beyond the heavy elements of the actinide group, α-decay and fission bring the Periodic Table to an end.

4.8 The contributions to B/A. Note that the surface, asymmetry and Coulomb terms all subtract from the bulk term.

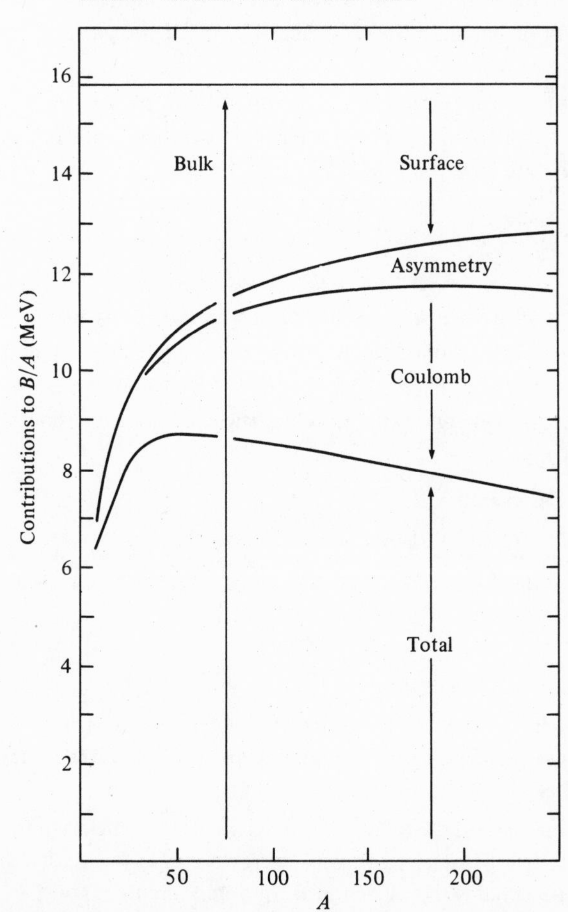

4.9 Nuclear binding and the nucleon–nucleon potential

To what extent do the nuclear properties discussed in this chapter follow from the nucleon–nucleon potential introduced in Chapter 3? Much theoretical effort has been expended on this question. In a nucleus containing three or more nucleons, the nuclear potential energy need not be the simple sum of two-body potentials over all pairs of nucleons: since the nucleons are composite particles, there may well be additional interactions.

Even if the possibility of additional interactions is not considered, the computations are not easy but it appears that the two-body potentials are the dominant contribution to the nuclear potential energy. For 'bulk' nuclear matter the Paris potential gives a value of 16 MeV/nucleon for the binding energy per nucleon, in good agreement with values found for the parameter a in the semi-empirical mass formula (4.5). However, the calculated density of nuclear matter is somewhat too high. The Paris potential gives 0.94 fm rather than the empirical 1.1 fm for the parameter in (4.2).

Similar semi-quantitative agreement is found when the two-nucleon potential is applied to particular light nuclei. For example, the binding energy of ^3_1H is calculated to be 7.38 MeV, and the experimental value (Table 4.2) is 8.48 MeV.

Problems

4.1 A relativistic electron whose rest mass can be neglected has energy E. It scatters elastically from a particle of mass M at rest and after the collision has turned through an angle θ and has energy E'.

(*a*) Show that the total energy of the struck particle after the collision is

$$E_M = E - E' + Mc^2.$$

(*b*) Show that its momentum is

$$P_M = [E^2 + E'^2 - 2EE' \cos\theta]^{\frac{1}{2}}/c.$$

(*c*) Hence (using $E_M^2 = P_M^2c^2 + M^2c^4$) show that the fraction of energy lost by the electron is

$$\frac{E-E'}{E} = \frac{1}{1 + Mc^2/[E(1-\cos\theta)]}.$$

For $E \sim$ a few hundred MeV, show that this is small if the struck particle is a heavy nucleus, and is large (except for $\theta \approx 0$) if the struck particle is an electron.

4.2 In quantum mechanics, the differential cross-section for the elastic scattering of a relativistic electron with energy $E \gg m_e c^2$ from a fixed electrostatic potential $\phi_c(r)$ is given in Born approximation, and neglecting the effects of electron spin, by

$$\frac{d\sigma}{d\Omega}=\left(\frac{E}{2\pi}\right)^2\left(\frac{1}{\hbar c}\right)^4\left(e\int\phi_c(r)e^{i\mathbf{q}\cdot\mathbf{r}}\,d^3\mathbf{r}\right)^2$$

where $\mathbf{q}$ is the difference between the final and the initial wave vectors of the electron.

(*a*) Show that $q=|\mathbf{q}|=(2E/\hbar c)\sin(\theta/2)$, where θ is the scattering angle.

(*b*) Poisson's equation relates the potential $\phi_c(r)$ to the charge density $e\rho_{ch}(r)$ by $\nabla^2\phi_c=-e\rho_{ch}/\varepsilon_0$. Noting $\nabla^2 e^{i\mathbf{q}\cdot\mathbf{r}}=-q^2e^{i\mathbf{q}\cdot\mathbf{r}}$, and integrating by parts, show that

$$\frac{d\sigma}{d\Omega}=\left(\frac{E}{2\pi}\right)^2\left(\frac{1}{\hbar c}\right)^4\frac{1}{q^4}\left(\frac{e^2}{\varepsilon_0}\int\rho_{ch}(r)e^{i\mathbf{q}\cdot\mathbf{r}}\,\mathbf{d}^3\mathbf{r}\right)^2.$$

For light nuclei (for which the Born approximation has a greater validity) a measured cross-section can be used to infer the Fourier transform of the charge distribution, as this example indicates.

4.3 Show that the characteristic velocity v of a lepton of mass m bound in an atomic orbit is given by $v/c\approx\hbar/amc=\frac{1}{137}$, where $a=(4\pi\varepsilon_0)\hbar^2/me^2$ is the appropriate Bohr radius for that lepton. Hence show that the muon mean life is long compared with the characteristic timescale a/v for its motion in an atomic orbit.

4.4 The ground state wave-function of a lepton of mass m in a Coulomb potential $-Ze^2/(4\pi\varepsilon_0 r)$ is

$$\psi(r)=\frac{1}{\sqrt{\pi}}\left(\frac{Z}{a}\right)^{\frac{3}{2}}e^{-Zr/a}$$

where $a=(4\pi\varepsilon_0)\hbar^2/me^2$, and the corresponding binding energy E is $Z^2\hbar^2/2ma^2$.

The finite size of the nucleus modifies the Coulomb energy for $r<R$, the nuclear radius, by adding a term of the approximate form

$$V(r)=-\frac{Ze^2}{4\pi\varepsilon_0 R}\left[\frac{3}{2}-\frac{r^2}{2R^2}-\frac{R}{r}\right].$$

(*a*) Show that the volume integral of this potential is

$$\int V(r)\,d^3\mathbf{r}=\frac{Ze^2R^2}{10\varepsilon_0}.$$

(*b*) Show that the first-order correction to the binding energy due to this term, $\Delta E=\int\psi^*(r)V(r)\psi(r)\,d^3\mathbf{r}$, is

$$\Delta E\approx\frac{e^2}{10\pi\varepsilon_0}\frac{Z^4R^2}{a^3}.$$

(Note that the lepton wave-function can be taken to be constant over nuclear dimensions.)

(*c*) For the nucleus $^{66}_{30}$Zn show that

$$\frac{\Delta E}{E} \approx 5 \times 10^{-6} \quad \text{for electrons.}$$

$$\frac{\Delta E}{E} \approx 0.2 \quad \text{for muons.}$$

4.5 Using Table 4.2 show that $^{8}_{4}$Be can decay to two α-particles with an energy release of 0.1 MeV, but that $^{12}_{6}$C cannot decay to three α-particles. Show that the energy released (including the energy of the photon) in the reaction $^{2}_{1}\text{H} + ^{4}_{2}\text{He} \to ^{6}_{3}\text{Li} + \gamma$ is 1.5 MeV.

4.6 Consider nuclei with small nucleon number A and such that $Z = N = A/2$. Neglecting the pairing term, show that the semi-empirical mass formula then gives the binding energy per nucleon

$$B/A = a - bA^{-\frac{1}{3}} - (d/4)A^{\frac{2}{3}}.$$

Show that this expression reaches a maximum for $Z = A/2 = 26$ (iron).

4.7 Using the formula (4.14) calculate Z for $A = 100$ and $A = 200$. Compare your results with Fig. 4.6 and comment.

4.8 The carbon isotope $^{14}_{6}$C is produced in nuclear reactions of cosmic rays in the atmosphere. It is β-unstable,

$$^{14}_{6}\text{C} \to ^{14}_{7}\text{N} + \text{e}^- + \bar{\nu}_e + 0.156 \text{ MeV},$$

with a mean life of 8270 years.

It is found that a gram of carbon, newly extracted from the atmosphere, gives on average 15.3 such radio-active decays per minute. What is the proportion of ^{14}C isotope in the carbon?

4.9 On the basis of the different properties of nuclei with even A and with odd A, explain why there are about 300 β-stable nuclei with masses up to that of $^{209}_{83}$Bi. What is the average number of isotopes per element?

5

Ground state properties of nuclei; the shell model

5.1 Nuclear potential wells

In the last chapter, we set out a semi-empirical theory for the binding energy of an atomic nucleus, and quantum-mechanical considerations came in only rather indirectly. Experimental atomic masses show deviations from the smooth curve given by the semi-empirical mass formula, deviations which we said were of quantum-mechanical origin. Since a nucleus in its ground state is a quantum system of finite size, it has angular momentum **J**, with quantum number j which is some integral multiple of $\frac{1}{2}$. If $j \neq 0$ the nucleus will have a magnetic dipole moment, and it may have an electric quadrupole moment as well.

The nuclear angular momentum and magnetic moment manifest themselves most immediately in atomic spectroscopy, where the interaction between the nuclear magnetic moment and the electron magnetic moments gives rise to the hyperfine structures of the electronic energy levels. In favourable cases both j and the magnetic moment may be deduced from this hyperfine splitting.

The observed values of nuclear angular momenta give strong support to the validity of a simple quantum-mechanical model of the nucleus: the nuclear shell model. In this model, each neutron moves independently in a common potential well that is the spherical average of the nuclear potential produced by all the other nucleons, and each proton moves independently in a common potential well that is the spherical average of the nuclear potential of all the other nucleons, together with the Coulomb potential of the other protons. Since the nuclear forces are of short range we can guess that the shape of such an average nuclear potential will reflect the nuclear density exhibited in Fig. 4.4. In particular it will be uniform in the central region and rise steeply in the surface region. Indeed, we can anticipate that the steep rise will be even more pronounced than the fall in nuclear density, since tunnelling into the classically-forbidden region will make the nuclear density near the surface more diffuse. Thus the potential wells for neutrons and protons will be, qualitatively, as sketched in Fig. 5.1(*a*) and Fig. 5.1(*b*).

In order to model the effect of proton charge, we have added to the proton well the Coulomb potential energy $U_c(r)$ of a proton in a sphere of uniform charge density and total charge $(Z-1)e$, corresponding to a uniform distribution of the other $(Z-1)$ protons. (Such a charge distribution gave an energy which agreed well with the empirical Coulomb contribution in the mass formula.) Elementary electrostatics gives

$$U_c(r)=\begin{cases}\dfrac{(Z-1)e^2}{4\pi\varepsilon_0 R}\left[\dfrac{3}{2}-\dfrac{r^2}{2R^2}\right], & r<R,\\[2ex] \dfrac{(Z-1)e^2}{4\pi\varepsilon_0 r}, & r>R,\end{cases}\tag{5.1}$$

where R is the nuclear radius.

Since the basic nucleon–nucleon interaction is state-dependent, there are other factors which affect the relative depths of the neutron and proton wells. For nuclei with more neutrons than protons the contribution of the strong nucleon–nucleon interaction to the potential is more attractive for the protons than for the neutrons, since a proton in such a nucleus is, on average, more subject to the neutron–proton interaction (recall the

5.1 A schematic representation of (*a*) the neutron potential well and (*b*) the proton potential well for the nucleus $^{208}_{82}$Pb. E_n^F and $(E_p^F-\bar{U})$ have been estimated using equation (5.5). The observed neutron separation energy S_n of 7.4 MeV implies a neutron well-depth of 51 MeV. The observed proton separation energy S_p of 8.0 MeV implies that $\bar{U}=11$ MeV. It will be seen that $E_n^F \approx E_p^F$.

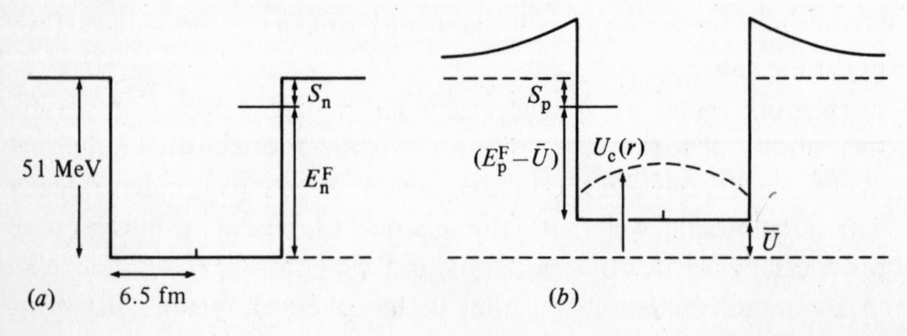

discussion of symmetry energy in § 4.5). This lowers the proton well-depth relative to $U_c(r)$, as is indicated in Fig. 5.1(*b*) (which is drawn with parameters appropriate to $^{208}_{82}$Pb). The nuclear shell model treats all these effects empirically.

5.2 Estimates of nucleon energies

Let us, for the moment, disregard the details of the nuclear potential at the surface, and replace it by a potential with infinitely high walls at $r=R$, which force the nucleon wave-functions to be zero at, and outside, $r=R$. We measure energies from the bottom of the neutron well, and for simplicity take the proton well to be raised with respect to the neutron well by a constant energy $\bar{U}$. This $\bar{U}$ represents the mean electrostatic potential and any asymmetry-contributions to the proton potential well. Then the Schrödinger equations for the neutron states ψ_n and proton states ψ_p are respectively

$$-\frac{\hbar^2}{2m_n}\nabla^2\psi_n=E_n\psi_n, \tag{5.2}$$

$$-\frac{\hbar^2}{2m_p}\nabla^2\psi_p=(E_p-\bar{U})\psi_p, \tag{5.3}$$

where any terms involving the intrinsic spin of the nucleons have been neglected.

Nucleons are fermions, and the Pauli exclusion principle requires that no two neutrons nor two protons are in the same state. Hence in the shell model of the ground state of a nucleus with N neutrons and Z protons, the lowest N neutron states are occupied up to some energy E_n^F, called the neutron Fermi energy, and the lowest Z proton states are occupied up to some energy E_p^F, the proton Fermi energy. To obtain a qualitative estimate of the energies involved, we suppose that N and Z are sufficiently large for us to use the elementary formula for the integrated density of states $\mathcal{N}(E)$, derived in Appendix B,

$$\mathcal{N}(E)=\frac{V}{3\pi^2}\left(\frac{2mE}{\hbar^2}\right)^{\frac{3}{2}} \tag{5.4}$$

where V is the volume of the system considered, in this case the nuclear volume $V=4\pi R^3/3$. Hence E_n^F and E_p^F are given by

$$N\approx\frac{V}{3\pi^2}\left(\frac{2m_nE_n^F}{\hbar^2}\right)^{\frac{3}{2}},\quad Z\approx\frac{V}{3\pi^2}\left(\frac{2m_p(E_p^F-\bar{U})}{\hbar^2}\right)^{\frac{3}{2}}, \tag{5.5}$$

since proton kinetic energies are given by $(E_p-\bar{U})$.

In lighter nuclei, $A\lesssim 40$, the number of neutrons and protons is approximately equal, so that N/V is close to half the 'density of nuclear matter' given by equation (4.2), i.e. $N/V \approx 0.085\,\text{fm}^{-3}$. For this density, equation (5.5) gives $E_n^F \approx 38$ MeV, irrespective of the particular nucleus. For heavier nuclei, this figure will increase somewhat, but for energies of this order the corresponding neutron velocities are quite low: $v_F^2/c^2 \approx 0.1$. This gives some justification for our use of the non-relativistic Schrödinger equation for the neutrons. Similarly, a proton at the Fermi energy E_p^F has kinetic energy $(E_p^F - \bar{U}) \approx 38$ MeV also.

Similar energies are to be expected in more realistic potentials of finite depth. In a finite well, the depth of E_n^F below the external potential outside the nucleus is equal to the energy required to detach a neutron from the nucleus. This energy, the *neutron separation energy* S_n, is given in terms of binding energies by

$$S_n(N, Z) = B(Z, N) - B(Z, N-1), \tag{5.6}$$

and hence is of the order of the binding energy per nucleon, about 8 MeV (Fig. 4.7, see also Table 4.2). Thus the total depth of the neutron well is ≈ 46 MeV.

The most stable nucleus of a given mass number $A = N + Z$ will have the neutron and proton Fermi energies approximately equal: if they were to differ greatly, the nucleus would be unstable to β-decay. A nucleon at the higher Fermi energy would decay to an empty state just above the lower Fermi energy, to form a new nucleus with different charge, and lower total energy.

The characteristic velocity of a nucleon at the Fermi energy, and the radius R of the nucleus, set a typical *nuclear time scale* t_{nuc}:

$$t_{nuc} = \frac{2R}{v_F} \approx 2.6 \times 10^{-23} \times A^{\frac{1}{3}}\ \text{s}. \tag{5.7}$$

5.3 Energy shells and angular momentum

To obtain more precise information about the nucleon levels we must solve the Schrödinger equations (5.2) and (5.3). For simplicity, we again take the potential wells with infinite walls. The Schrödinger equations are separable in (r, θ, ϕ) coordinates so that

$$\psi(r, \theta, \phi) = u_l(r) Y_{lm}(\theta, \phi),$$

where $l = 0, 1, 2, 3, \ldots$ and $m = -l, -l+1, \ldots, l-1, l$, are orbital angular momentum quantum numbers (Appendix C). States with $l = 0, 1, 2, 3, 4, \ldots$ are called s, p, d, f, g, h, ... states, the notation having been established in the early days of atomic spectroscopy. Taking equation (5.2) for neutrons (and the equation for protons differs only in the shift $\bar{U}$ in energy), the radial function $u_l(r)$ satisfies

$$\frac{-\hbar^2}{2m_n}\frac{1}{r}\frac{d^2}{dr^2}(ru_l)+\frac{\hbar^2}{2m_n}\frac{l(l+1)}{r^2}u_l=Eu_l, \tag{5.8}$$

with the boundary conditions that $u(r)$ is finite at $r=0$ and zero at $r=R$.

When $l=0$ (s-states) we see immediately that the solutions finite at $r=0$ are

$$u_0(r)=\frac{\sin(kr)}{kr} \quad \text{with } E=\frac{\hbar^2k^2}{2m_n}. \tag{5.9}$$

The boundary condition at $r=R$ is satisfied by taking $k=k_n=n\pi/R$, where $n=1, 2, 3, \ldots$, and the corresponding energy levels 1s, 2s, 3s, ..., are given by

$$E(ns)=\frac{\hbar^2}{2m_n}\left(\frac{n\pi}{R}\right)^2. \tag{5.10}$$

When $l=1$ (p-states) the Schrödinger equation is

$$-\frac{\hbar^2}{2m_n}\frac{1}{r}\frac{d^2}{dr^2}(ru_1)+\frac{\hbar^2}{m_n r^2}u_1=Eu_1, \tag{5.11}$$

and it is straightforward to check by differentiation that the solution finite at $r=0$ is

$$u_1(r)=\frac{\sin(kr)}{(kr)^2}-\frac{\cos(kr)}{(kr)}, \; E=\frac{\hbar^2}{2m_n}k^2.$$

We must again choose k so that $u_1(R)=0$. Let us write $kR=x$. The values of x for which $u_1(R)=0$ are $x_{1p}=4.49$, $x_{2p}=7.73, \ldots$, and the corresponding energies are $E(np)=(\hbar^2/2m_n)(x_{np}/R)^2$. It should be noted that the labelling of energy levels, $n=1, 2, 3, \ldots$, for each l value, differs from that conventionally adopted in atomic physics.

In fact $u_0(r)$ and $u_1(r)$ are special cases of the *spherical Bessel functions* $j_l(kr)$. For arbitrary l the zeros of $j_l(x)$, which give the allowed values of k, are tabulated in standard tables of mathematical functions, and given in column (2) of Table 5.1. Thus for any l the levels $E(n, l)=(\hbar^2/2m)(x_{nl}/R)^2$ are easily determined, and hence their sequence in order of increasing energy. This is given in columns (1) and (7) of Table 5.1. For each $E(n, l)$, there are $(2l+1)$ allowed values of the quantum number m. Since we have so far neglected any coupling between the intrinsic spin of a nucleon and its orbital motion, each nucleon has two possible spin states, which may be characterised by $m_s=\frac{1}{2}$, $m_s=-\frac{1}{2}$, so that there are $(4l+2)$ states of the same energy for a given (n, l).

The sequence of the levels $E(n, l)$ is not very sensitive to the precise details

Table 5.1

		Neutron		Proton		
	x_{nl}					
1s	3.14	$1s_{\frac{1}{2}}$	2	2	$1s_{\frac{1}{2}}$	1s
1p	4.49	$1p_{\frac{3}{2}}$	6	6	$1p_{\frac{3}{2}}$	1p
		$1p_{\frac{1}{2}}$	8	8	$1p_{\frac{1}{2}}$	
1d	5.76	$1d_{\frac{5}{2}}$	14	14	$1d_{\frac{5}{2}}$	1d
		$2s_{\frac{1}{2}}$	16	16	$2s_{\frac{1}{2}}$	
2s	6.28	$1d_{\frac{3}{2}}$	20			2s
				20	$1d_{\frac{3}{2}}$	
1f	6.99	$1f_{\frac{7}{2}}$	28			
		$2p_{\frac{3}{2}}$	32	28	$1f_{\frac{7}{2}}$	
2p	7.73	$1f_{\frac{5}{2}}$	38			1f
		$2p_{\frac{1}{2}}$	40	32	$2p_{\frac{3}{2}}$	
1g	8.18	$1g_{\frac{9}{2}}$	50	38	$1f_{\frac{5}{2}}$	2p
		$2d_{\frac{5}{2}}$	56	40	$2p_{\frac{1}{2}}$	
2d	9.10	$1g_{\frac{7}{2}}$	64			
		$1h_{\frac{11}{2}}$	76	50	$1g_{\frac{9}{2}}$	
1h	9.36	$3s_{\frac{1}{2}}$	78			1g
		$2d_{\frac{3}{2}}$	82	58	$1g_{\frac{7}{2}}$	
3s	9.42	$2f_{\frac{7}{2}}$	90	64	$2d_{\frac{5}{2}}$	
		$1h_{\frac{9}{2}}$	100			
2f	10.42	$3p_{\frac{3}{2}}$	104			2d
1i	10.51	$1i_{\frac{13}{2}}$	118	76	$1h_{\frac{11}{2}}$	
		$2f_{\frac{5}{2}}$	124	80	$2d_{\frac{3}{2}}$	1h
3p	10.90	$3p_{\frac{1}{2}}$	126	82	$3s_{\frac{1}{2}}$	
						3s
		$2g_{\frac{9}{2}}$	136			
1j	11.66	$1i_{\frac{11}{2}}$	148	92	$1h_{\frac{9}{2}}$	
		$2g_{\frac{7}{2}}$	156	100	$2f_{\frac{7}{2}}$	
2g	11.70					

The first and last columns give the sequence of energy levels in a spherical well with infinite walls. The second column gives the corresponding values of $x_{nl} = k_{nl}R$. The third column gives the observed sequence of spin-orbit coupled levels for neutrons, and the fourth the cumulative number of available states in these levels. The remaining two columns give the levels and number of states for protons. The spacings are chosen so that the filling of the neutron and proton shells for stable nuclei is approximately in step down the columns. Lines are drawn at the 'magic numbers'.

of the well, and is much the same for a well of finite depth and appropriately rounded shape. If N neutrons are put into the neutron well, the ground state (which may be degenerate) will correspond to the occupation of the N lowest lying energy states. Figure 5.2 expresses graphically the number of states available in terms of the dimensionless quantity $x = kR$. $\mathcal{N}(x)$ is the number of states with energy less than $(\hbar^2/2m_n)(x/R)^2$ – or, equivalently, the number of zeros x_{nl} with $x_{nl} < x$, each zero being counted $(4l+2)$ times. Also drawn is the asymptotic formula, valid for large x,

$$\mathcal{N}(x) = \frac{4x^3}{9\pi}\left(1 - \frac{9\pi}{8x}\right) = \frac{V}{3\pi^2}\left(\frac{2m_n E}{\hbar^2}\right)^{\frac{3}{2}}\left[1 - \frac{3\pi}{8}\frac{S}{V}\left(\frac{\hbar^2}{2m_n E}\right)^{\frac{1}{2}}\right]. \quad (5.13)$$

This is an extension of the usual density of states formula we used in equation (5.4) and includes a correction for the effects of the surface (of area S). It follows the exact $\mathcal{N}(x)$ remarkably closely.

So far we have neglected spin. The crucial step in establishing the nuclear shell model was the recognition that there must also be a *spin-orbit coupling* term in the self-consistent potential seen by the nucleons, of the form

$$U_{so}(r)\,\mathbf{L}\cdot\mathbf{s}. \quad (5.14)$$

5.2 The exact $\mathcal{N}(x)$ compared with the asymptotic formula of equation (5.13). (See Baltes, H. P. and Hilf, E. R. (1976), *Spectra of Finite Systems*, for the derivation of (5.13).)

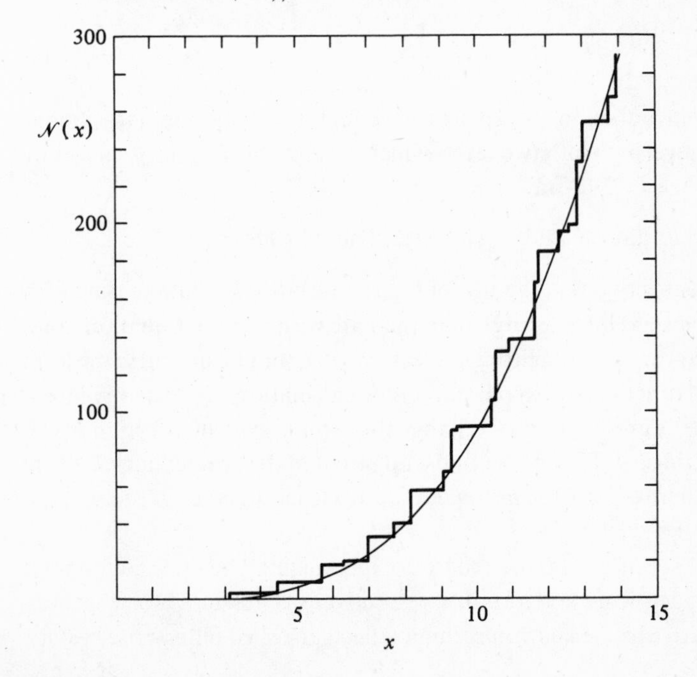

A term like this is not perhaps too surprising since in the basic nucleon–nucleon interaction, equation (3.4), there is coupling between the spins and the orbital motions of the nucleons.

With the introduction of this spin-orbit coupling term into the potential, $\mathbf{L}^2$ and $\mathbf{s}^2$ are still conserved since

$$[\mathbf{L}^2, \mathbf{L}\cdot\mathbf{s}]=0, \quad [\mathbf{s}^2, \mathbf{L}\cdot\mathbf{s}]=0, \tag{5.15}$$

so that l and $s(=\frac{1}{2})$ remain good quantum numbers, but m and m_s are 'good quantum numbers' no longer, since

$$[L_z, \mathbf{L}\cdot\mathbf{s}]\neq 0, \quad [m_s, \mathbf{L}\cdot\mathbf{s}]\neq 0. \tag{5.16}$$

Thus the magnitudes, but not the directions, of $\mathbf{L}$ and $\mathbf{s}$ are preserved. However, the *total angular momentum* $\mathbf{J}=\mathbf{L}+\mathbf{s}$ is of course conserved, so that states may be specified by the quantum numbers (l, s, j, j_z). In Appendix C it is shown that for a given l and $s=\frac{1}{2}$, the allowed values of j are: $l+\frac{1}{2}$, with $[2(l+\frac{1}{2})+1]=2l+2$ allowed values of j_z; and $l-\frac{1}{2}$, with $[2(l-\frac{1}{2})+1]=2l$ allowed values of j_z. The parity of the state specified by $(l, s, l\pm\frac{1}{2}, j_z)$ is $(-1)^l$.

The expectation value of $\mathbf{L}\cdot\mathbf{s}$ may be obtained from the identity

$$\mathbf{L}\cdot\mathbf{s}=\tfrac{1}{2}[(\mathbf{L}+\mathbf{s})^2-\mathbf{L}^2-\mathbf{s}^2]=\tfrac{1}{2}[\mathbf{J}^2-\mathbf{L}^2-\mathbf{s}^2], \tag{5.17}$$

so that

$$\begin{aligned}(l, s, j, j_z|\mathbf{L}\cdot\mathbf{s}|l, s, j, j_z) &= \tfrac{1}{2}[j(j+1)-l(l+1)-s(s+1)]\hbar^2 \\ &= \begin{cases} \tfrac{1}{2}l\hbar^2 & \text{if } j=l+\tfrac{1}{2} \\ -\tfrac{1}{2}(l+1)\hbar^2 & \text{if } j=l-\tfrac{1}{2} \end{cases}\end{aligned} \tag{5.18}$$

Thus the introduction of spin-orbit coupling splits the $(4l+2)$-fold degenerate level (n, l) into two levels which we may label by $nl_{l+\frac{1}{2}}, nl_{l-\frac{1}{2}}$. For example, when $l=2$ (d-states),

$$\text{nd (10 states)} \rightarrow \text{nd}_{\frac{5}{2}} \text{ (6 states) and nd}_{\frac{3}{2}} \text{ (4 states).}$$

Experiment shows that the sign of $U_{so}(r)$ is negative so that the state with $j=l+\frac{1}{2}$ always has lower energy than the state with $j=l-\frac{1}{2}$. Equation (5.18) suggests that the energy splitting increases with l, though of course the form of the radial function is also relevant in the calculation of the energy levels. The splitting is great enough to change the orbital sequence of columns (1) and (7) of Table 5.1. This effect is most apparent in the heavier nuclei, where because of increasing R the orbital levels are closer together in energy than in the lighter nuclei.

The sequence of 'shells' inferred from experiment is shown in columns (3) and (6) for neutrons and protons respectively. The shift of the proton column relative to the neutron column reflects the shell filling observed for the β-stable nuclei. In the filling of these levels there are departures from the order given in the case of a few particular nuclei.

The major success of the shell model is the prediction of the angular momenta of nuclei in their ground states. These values follow simple rules: nuclei with an even number of protons and an even number of neutrons (even–even nuclei) have angular momentum zero and positive parity, nuclei with an even number of protons and an odd number of neutrons or vice-versa (even–odd nuclei), have angular momentum and parity equal to that of the odd nucleon in the shell that is being filled. We saw in Chapter 4 that it was energetically favourable for nuclei to contain even numbers of protons and even numbers of neutrons. The information from nuclear spins makes more precise the origin of this energy: it seems that it is energetically advantageous for nuclei to take pairs of protons and pairs of neutrons into the energy shells, with the angular momenta of the pairs coupled to zero, $\mathbf{J}_1+\mathbf{J}_2=0$, so that the angular momentum and parity of an unpaired nucleon is the angular momentum and parity of the whole nucleus. There are some exceptions to this last rule, but remarkably few considering its simplicity.

In the case of odd–odd nuclei, the odd proton and odd neutron do not combine their angular momenta in any systematic way; there is no very clear empirical rule, and no simple theory. Indeed, odd–odd nuclei are altogether energetically disfavoured. There are only four stable odd–odd nuclei ($^2_1$H, $^6_3$Li, $^{10}_5$B, $^{14}_7$N), the rest undergo β-decay to become even–even (Chapter 4).

The rules may be seen obeyed by the light nuclei of Table 4.2. For example $^{17}_8$O has one odd neutron in the $1d_{\frac{5}{2}}$ shell, and spin and parity $\frac{5}{2}^+$.

5.4 Magic numbers

In Table 5.1 lines have been drawn where the total number of states in the shells above the line is 2, 8, 20, 28, 50, 82 and 126, the so-called 'magic numbers'. The first two numbers just correspond to the filling of 1s and 1p shells. The others appear to be somewhat arbitrary, but it is found empirically that the energy gaps to the next shell are greater than average at these points in the sequence. Nuclei having Z or N equal to one of these numbers have properties which reflect the existence of such a gap. For example, tin ($Z=50$) has ten stable isotopes, and there are seven stable elements having $N=82$ (see Fig. 4.6). These examples illustrate that there must be a large energy gap at $Z=50$ in the proton sequence and at $N=82$ in the neutron sequence, as is clear from our discussion in § 5.2 of the condition $E_n^F \approx E_p^F$.

The departures from the smooth curve of binding energy per nucleon, given by the semi-empirical mass formula, are associated with the nuclear shell structure and the magic numbers. Magic number nuclei are particularly strongly bound and have been marked on Fig. 4.7. The heaviest α-stable nuclei are ${}^{208}_{82}\mathrm{Pb}$, which is 'doubly magic' with $N=126$, $Z=82$, and ${}^{209}_{83}\mathrm{Bi}$ which has $N=126$.

The magic numbers were identified experimentally before the shell model was established, and indeed the indications of shell structure provided by the existence of these numbers was a strong motivation for the formulation of the shell model. Other consequences of the magic numbers will be mentioned in the sections on the excited states of nuclei and atomic abundances in the Solar System.

5.5 The magnetic dipole moment of the nucleus

The successful description of nuclear angular momentum indicates the essential validity of the shell model. The model also gives a qualitative understanding of the magnetic dipole moments of nuclei. The magnetic moments of paired nucleons, like their spins, cancel exactly, and all even–even nuclei are found to have zero magnetic moments. A nucleus with angular momentum operator **J** (quantum numbers j, j_z) has a magnetic moment operator $\boldsymbol{\mu}$ which, averaged over the nucleons, must be aligned with **J**, since **J** is the only vector available giving a preferred direction. The *magnetic dipole moment* μ is defined by writing

$$\langle\boldsymbol{\mu}\rangle=\frac{\mu}{(j\hbar)}\langle\mathbf{J}\rangle, \tag{5.19}$$

where the brackets $\langle\cdots\rangle$ indicate any matrix element between the $(2j+1)$ states labelled by j_z.

In a magnetic field $\mathbf{B}=(0,0,B)$, which specifies the z-direction, the magnetic potential energy of the nucleus in the field is the expectation value of $-\boldsymbol{\mu}\cdot\mathbf{B}=-\mu_z B$. For a state of given j_z, this energy is, from equation (5.19),

$$E(j_z)=-\mu(j_z/j)B \tag{5.20}$$

so that there are $(2j+1)$ equally spaced energy levels corresponding to $j_z=-j, -j+1, \ldots, j$.

Transitions between these levels may be induced by a radio-frequency oscillating electromagnetic field of angular frequency ω where

$$\hbar\omega=|\mu|B/j. \tag{5.21}$$

Measurements of this resonance frequency in a known magnetic field give a precise value for the magnetic dipole moment. The phenomenon is called *nuclear magnetic resonance*, and has many applications in physics, chemistry and biology.

5.6 Calculation of the magnetic dipole moment

In our simple version of the shell model, the magnetic moment of an odd-A nucleus will arise entirely from the unpaired nucleon. If this unpaired nucleon is a proton, its orbital motion will give, as in classical magnetism, a moment

$$\boldsymbol{\mu}_L = \frac{e\mathbf{L}}{2m_\mathrm{p}} = \mu_N\left(\frac{\mathbf{L}}{\hbar}\right) \tag{5.22}$$

where $\mu_N = e\hbar/2m_\mathrm{p}$ is the *nuclear magneton.* A neutron, since it is uncharged, will give no orbital contribution.

To this must be added the intrinsic magnetic moment of the nucleon,

$$\boldsymbol{\mu}_\mathrm{s} = g_\mathrm{s}\mu_N\left(\frac{\mathbf{s}}{\hbar}\right), \tag{5.23}$$

where (using the values quoted in equation (3.2)), $g_\mathrm{s} = 5.59$ for a proton and $g_\mathrm{s} = -3.83$ for a neutron.

Thus the total magnetic moment operator for a single nucleon is

$$\boldsymbol{\mu} = \boldsymbol{\mu}_L + \boldsymbol{\mu}_\mathrm{s} = \mu_N[g_L\mathbf{L} + g_\mathrm{s}\mathbf{s}]/\hbar, \tag{5.24}$$

where $g_L = 1$ for a proton and $g_L = 0$ for a neutron. We can write this as

$$\boldsymbol{\mu} = \mu_N[\tfrac{1}{2}(g_L + g_\mathrm{s})(\mathbf{L} + \mathbf{s}) + \tfrac{1}{2}(g_L - g_\mathrm{s})(\mathbf{L} - \mathbf{s})]/\hbar,$$

and take the scalar product with $\mathbf{J} = \mathbf{L} + \mathbf{s}$ to give

$$\boldsymbol{\mu}\cdot\mathbf{J} = \mu_N[\tfrac{1}{2}(g_L + g_\mathrm{s})\mathbf{J}^2 + \tfrac{1}{2}(g_L - g_\mathrm{s})(\mathbf{L}^2 - \mathbf{s}^2)]/\hbar$$

(since $\mathbf{L}$ and $\mathbf{s}$ commute). Then the expectation value of each side of this equation for a state specified by (j, j_z, l, s) gives, using equation (5.19)

$$\left(\frac{\mu}{j}\right)j(j+1) = \mu_N[\tfrac{1}{2}(g_L + g_\mathrm{s})j(j+1) + \tfrac{1}{2}(g_L - g_\mathrm{s})(l(l+1) - s(s+1))]$$

so that

$$\mu = \mu_N\left[\frac{1}{2}(g_L + g_\mathrm{s})j + \frac{1}{2}(g_L - g_\mathrm{s})\frac{(l-s)(l+s+1)}{(j+1)}\right]. \tag{5.25}$$

Since $s = \frac{1}{2}$ and $j = l \pm \frac{1}{2}$ we finally obtain for the contribution from the unpaired nucleon

$$\left.\begin{aligned} \mu &= \mu_N[jg_L - \tfrac{1}{2}(g_L - g_\mathrm{s})] && \text{for } j = l + \tfrac{1}{2} \\ \mu &= \mu_N\left[jg_L + \frac{j}{2(j+1)}(g_L - g_\mathrm{s})\right] && \text{for } j = l - \tfrac{1}{2} \end{aligned}\right\}, \tag{5.26}$$

which are referred to as the 'Schmidt values'.

These predictions of the simple model for a nucleus with an odd unpaired nucleon are not grossly wrong: almost all the observed magnetic moments for such nuclei lie between these two values. But they are not accurate predictions and there is no generally accepted explanation of the discrepancies. One possible reason is that the intrinsic magnetic moment of the nucleon is smaller in a nuclear environment than in free space. Another interpretation is that the magnetic moments provide a more sensitive test of the nuclear shell model than does the nuclear spin, and cooperative effects which we have neglected may contribute.

5.7 The electric quadrupole moment of the nucleus

Nuclei with spin $\geqslant 1$ usually have small permanent electric quadrupole moments. The size of this electric quadrupole moment gives an indication of the extent to which the distribution of charge (and hence matter) in the nucleus deviates from spherical symmetry. A nucleus is coupled through its electric quadrupole moment to the gradient, at the nuclear site, of the external electric field produced by the molecular or crystalline environment of the nucleus. Like the nuclear magnetic dipole moment, the nuclear electric quadrupole moment provides a sensitive probe of this environment for chemistry and condensed matter physics.

Classically, the energy of a nuclear charge distribution $e\rho_{\text{ch}}(\mathbf{r})$ in an external electrostatic potential $\phi(\mathbf{r})$ is

$$U = e\int \rho_{\text{ch}}(\mathbf{r})\phi(\mathbf{r})\,\mathrm{d}^3\mathbf{r}.$$

We take the origin $\mathbf{r}=0$ to be the centre of mass of the nucleus. Since $\rho_{\text{ch}}(\mathbf{r})$ is confined to the small nuclear volume we can approximate $\phi(\mathbf{r})$ by the first few terms of a Taylor series,

$$\phi(\mathbf{r}) \approx \phi(0) - \mathbf{r}\cdot\mathbf{E} + \frac{1}{2}\sum_{i,j} x_i x_j \phi_{ij}, \tag{5.27}$$

where $\mathbf{E} = -\nabla\phi$ is the electric field, and

$$\phi_{ij} = \frac{\partial^2\phi}{\partial x_i\,\partial x_j} = -\frac{\partial E_i}{\partial x_j}, \quad \text{all evaluated at } \mathbf{r}=0.$$

Here the indices i, j run from one to three, and we are using the notation $\mathbf{r} = (x_1, x_2, x_3)$, etc.

We then have

$$\mathrm{U} = eZ\phi(0) - \mathbf{E}\cdot\mathbf{d} + \tfrac{1}{2}e\sum_{i,j}\phi_{ij}\int \rho_{\text{ch}}(\mathbf{r})x_i x_j\,\mathrm{d}^3\mathbf{r},$$

where $\mathbf{d} = e\int \rho_{\text{ch}}(\mathbf{r})\mathbf{r}\,\mathrm{d}^3\mathbf{r}$ is the electric dipole moment. The first term would be the energy if the nuclear charge Ze were a point charge at the origin. The second term is the electric dipole energy. Apart from negligible weak interaction effects, nuclear charge densities have the reflection symmetry $\rho_{ch}(\mathbf{r}) = \rho_{ch}(-\mathbf{r})$; thus nuclear electric dipole moments are zero, and the effects of the extended nuclear charge distribution appear in the term

$$\Delta U = \tfrac{1}{2}e \sum_{i,j} \phi_{ij} \int \rho_{ch}(\mathbf{r}) x_i x_j \, d^3\mathbf{r},$$

which in general does not vanish.

If we neglect the charge density of atomic electrons at the nucleus, the external potential ϕ satisfies Laplace's equation

$$\nabla^2 \phi = \sum_i \phi_{ii} = 0. \tag{5.28}$$

We can therefore re-write ΔU in the form

$$\Delta U = \frac{e}{6} \sum_{i,j} \phi_{ij} Q_{ij},$$

where

$$Q_{ij} = \int \rho_{ch}(\mathbf{r})[3x_i x_j - \mathbf{r}^2 \delta_{ij}] \, d^3\mathbf{r}.$$

Q_{ij} is defined to be the *quadrupole moment tensor* of the classical charge distribution. δ_{ij} is the usual Kronecker δ. The additional term containing δ_{ij} does not change ΔU, because of (5.28), but makes $Q_{ij} \equiv 0$ if $\rho_{ch}(\mathbf{r})$ is spherically symmetric.

Dimensional analysis suggests that typically

$$\Delta U \sim \frac{e^2(\text{nuclear dimension})^2}{4\pi\varepsilon_0(\text{atomic dimension})^3} \sim 10^{-9} \text{ eV}.$$

Such small energy shifts are detectable in radio-frequency spectroscopy, but it is clear that higher multi-pole moments arising from further terms in the Taylor expansion (5.27) will be unimportant.

The charge distribution in a nucleus must of course be treated quantum mechanically rather than classically, and we define the *electric quadrupole moment* operator by

$$Q_{ij} = \sum_{\text{protons } p} [3x_{pi}x_{pj} - \delta_{ij}\mathbf{r}_p^2]$$

where now the x_{pi} are the proton coordinates.

Just as the matrix elements of the (vector) magnetic dipole operator $\boldsymbol{\mu}$ are proportional to the matrix elements of the (vector) angular momentum operator $\mathbf{J}$, it can be shown that for the tensor operator Q_{ij}

$$\langle Q_{ij} \rangle = C \langle [\tfrac{3}{2}(J_i J_j + J_j J_i) - \delta_{ij}\mathbf{J}^2] \rangle,$$

where C is a constant, and again the brackets indicate any matrix elements between the $(2j+1)$ nuclear states of angular momentum quantum number j, labelled by j_z. Thus all the matrix elements of Q_{ij} are determined by a single quantity. It is conventional to take the expectation value of Q_{33} in the state with j_z equal to its maximum value j, and define this as the *nuclear electric quadrupole moment* Q, so that

$$Q = C[3j^2 - j(j+1)] = Cj(2j-1),$$

or

$$C = \frac{Q}{j(2j-1)}.$$

All other matrix elements are then determined in terms of Q. Note that Q vanishes for nuclei with $j=0$ or $j=\frac{1}{2}$.

Experimental values of Q are obtained from spectral measurements on systems in which the field gradients can be accurately calculated. These values are often very much larger than, and sometimes differ in sign from, the predictions of the simple shell model. The implication is that the deformation of many nuclei from spherical symmetry is much larger than would be expected from the simple independent particle shell model. In reality the deformations must result from collective effects involving several nucleons. As might be expected, deviations from spherical symmetry are least in the neighbourhood of closed shells and largest for nuclei with shells which are around half-full.

Problems

5.1 In the model of § 5.2, verify that when $N=Z=A/2$ then $E_n^F = E_p^F - \bar{U} \approx$ 38 MeV.

5.2(*a*) Show that in the model of § 5.2 the total kinetic energy of a nucleus containing N neutrons and Z protons is

$$[\tfrac{3}{5}NE_n^F + \tfrac{3}{5}Z(E_p^F - \bar{U})].$$

(*b*) For $(N-Z) \ll A$, this expression may be Taylor expanded about $N_0 = A/2$, $Z_0 = A/2$, $E_0^F \approx 38$ MeV. Show that

$$E_n^F \approx E_0^F\left[1 + \frac{2}{3}\frac{\Delta N}{N_0} - \frac{1}{9}\left(\frac{\Delta N}{N_0}\right)^2\right],$$

$$(E_p^F - \bar{U}) = E_0^F\left[1 - \frac{2}{3}\frac{\Delta N}{N_0} - \frac{1}{9}\left(\frac{\Delta N}{N_0}\right)^2\right],$$

where $\Delta N = -\Delta Z = (N-Z)/2$.

(*c*) Hence show that the total kinetic energy of the nucleons in the nucleus is approximately

$$\tfrac{3}{5}E_0^F A + \tfrac{1}{3}E_0^F \frac{(N-Z)^2}{A}$$

and therefore contributes ≈ -23 MeV to the coefficient 'a' in the semi-empirical mass formula, and ≈ 13 MeV to the symmetry coefficient 's' (equation (4.5)).

5.3(*a*) Show from equation (5.1) that the average Coulomb energy of a proton in a nucleus of atomic number Z is

$$\bar{U}_c = \frac{6}{5}\frac{(Z-1)e^2}{4\pi\varepsilon_0 R}.$$

(*b*) Show that for ${}^{208}_{82}\mathrm{Pb}$, $\bar{U}_c \approx 2\bar{U}$.

5.4 ${}^{40}_{20}\mathrm{Ca}$ is the heaviest stable nucleus with $Z=N$. (It is doubly magic.) The neutron separation energy is 15.6 MeV. Estimate the proton separation energy, and compare your estimate with the empirical value of 8.3 MeV.

5.5 Suggest values for the spins and parities of the following nuclei in their ground states:

${}^{31}_{15}\mathrm{P}$, ${}^{67}_{30}\mathrm{Zn}$, ${}^{115}_{49}\mathrm{In}$.

5.6 The measured spins, parities and magnetic moments of some nuclei are:

$${}^{43}_{20}\mathrm{Ca}\left(\frac{7^-}{2}, -1.32\mu_N\right), \quad {}^{93}_{41}\mathrm{Nb}\left(\frac{9^+}{2}, 6.17\mu_N\right), \quad {}^{137}_{56}\mathrm{Ba}\left(\frac{3^+}{2}, 0.931\mu_N\right),$$

$${}^{197}_{79}\mathrm{Au}\left(\frac{3^+}{2}, 0.145\mu_N\right), \quad {}^{26}_{13}\mathrm{Al}(5^+, \text{not known}).$$

Compare these values with the predictions of the shell model.

5.7 Calculate the nuclear magnetic resonance frequencies for (*a*) protons, (*b*) ${}^{43}_{20}\mathrm{Ca}$ (see Problem 5.6), in a magnetic field of 1 tesla ($=10^4$ gauss).

5.8 Show that for a uniformly charged ellipsoid of revolution

$$\frac{x^2+y^2}{a^2}+\frac{z^2}{b^2}\leqslant 1,$$

of total charge Ze,

$$Q_{zz}=\tfrac{2}{5}Z(b^2-a^2).$$

The nucleus ${}^{176}_{71}\mathrm{Lu}$ has $j=7$ and a very large electric quadrupole moment of 8.0 barns. Suppose the nucleus in the state with $j_z=7$ has approximately an ellipsoidal charge distribution of the form above. Calculate a and b.

6

Alpha decay and spontaneous fission

6.1 Energy release in α-decay

We saw at the end of Chapter 4 that the binding energy per nucleon curve of the β-stable nuclei has a maximum in the neighbourhood of iron (Fig. 4.7), and that the heavier elements may be unstable to spontaneous disintegration. The principal mode of break-up is by emission of a ${}^4_2$He nucleus. Historically, the particles emitted in the decays of naturally occurring α-unstable nuclei were called α-particles before they were identified by Rutherford in 1908 as ${}^4_2$He nuclei, and the name has stayed.

The kinetic energy release $Q(A, Z)$ in an α-decay of a nucleus (A, Z) is given in terms of the binding energies of the parent and daughter nuclei by

$$Q(A, Z) = B(A-4, Z-2) + 28.3 \text{ MeV} - B(A, Z), \tag{6.1}$$

(where 28.3 MeV is the experimental binding energy of the ${}^4_2$He nucleus). If the nucleus is assumed to lie on the β-stability curve, given approximately by equation (4.14), then Q may be calculated as a function of Z (or of A) from the semi-empirical mass formula, using equation (4.5). Neglecting the pairing energy term, the effects of which are small, the resulting smooth $Q(Z)$ is plotted in Fig. 6.1 for $Z > 50$. Negative values of Q imply absolute stability against α-decay. Also shown are Q values calculated from the experimentally measured masses of β-stable nuclei and their corresponding daughter nuclei. The trend of the experimental points is given correctly by the semi-empirical formula. Though the detailed predictions can be out by as much as 5 MeV, this is very small (<0.3%) compared with the total binding energies of the nuclei in this region. The main deviation from the simple formula is due to the extra binding energy of nuclei around the double-closed shell nucleus $^{208}_{82}Pb$. This extra binding energy not only makes these nuclei more nearly stable than average, but also makes less stable the nuclei immediately above them. At higher Z, around $^{238}_{92}U$, there is another small region of relative stability.

From Fig. 6.1 it will be seen that β-stable nuclei with $Z>66$ (and a few with $Z\leqslant 66$) are in principle unstable to α-decay. In practice the decay rate is so low as to be almost unobservable if the energy release Q is <4 MeV. Up to Bi ($Z=83$) the lifetimes of β-stable nuclei are many orders of magnitude greater than the age of the Earth.

6.2 The theory of α-decay

It is the electrostatic force which is responsible for inhibiting the α-decay of those nuclei for which the decay is energetically favourable. As an

6.1 Experimental and theoretical α-decay energies $Q=B(A-4, Z-2)+28.3\text{ MeV}-B(A, Z)$, as a function of the atomic number Z of the parent nucleus. The experimental points are from cases where both parent nucleus and daughter nucleus are β-stable. The theoretical curve is from equation (4.5) (neglecting the pairing energies) together with equation (4.14).

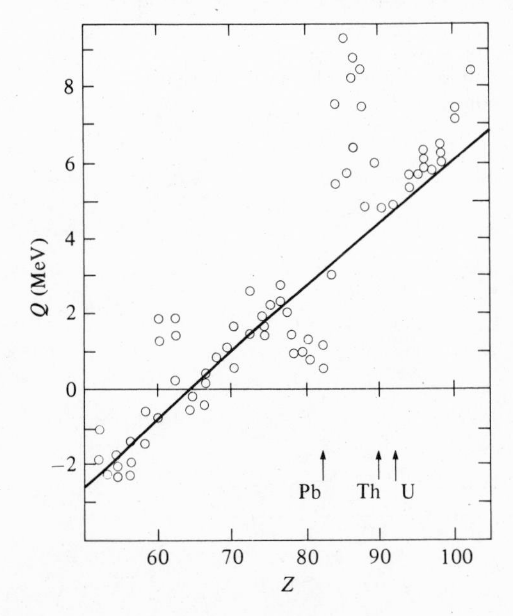

example, consider the decay of bismuth to thallium,

$$^{209}_{83}\text{Bi} \rightarrow ^{205}_{81}\text{Tl} + ^{4}_{2}\text{He} + 3.11\ \text{MeV},$$

which is in principle possible but is not observed.

Figure 6.2 shows the electrostatic potential energy of the α-particle (charge $2e$) at a distance r from a thallium nucleus ($Z_{\text{Tl}} = 81$), $2Z_{\text{Tl}}e^2/(4\pi\varepsilon_0)r$. Also indicated on the graph is the distance r_s at which the strong interaction with the thallium nucleus takes over, which we estimate as

$$r_s = 1.1[(205)^{\frac{1}{3}} + 4^{\frac{1}{3}}]\ \text{fm} = 8.23\ \text{fm},$$

where we have used equation (4.2) for the radii of the Tl and He nuclei. It is around this distance, in the surface region of the parent Bi nucleus, that the α-particle can be considered to be formed.

The graph immediately explains why the α-particle finds it difficult to escape even if it is formed. At r_s the height of the Coulomb potential is 28.4 MeV, very much greater than the energy $Q = 3.11$ MeV of the α-particle. Classically, it cannot penetrate the barrier and is free to move only at distances greater than r_c where r_c is given by

$$Q = \frac{2Z_{\text{Tl}}e^2}{(4\pi\varepsilon_0)r_c}; \tag{6.2}$$

r_c is the classical distance of closest approach to the nucleus of an α-particle of energy Q coming from the outside. For thallium, $r_c = 75$ fm. Thus

6.2 The potential energy of an α-particle in the Coulomb field of a thallium nucleus, as a function of the separation distance r. At $r \approx r_s$ the α-particle from the decay of bismuth is formed. At $r = r_c$ it has penetrated through the classically forbidden region.

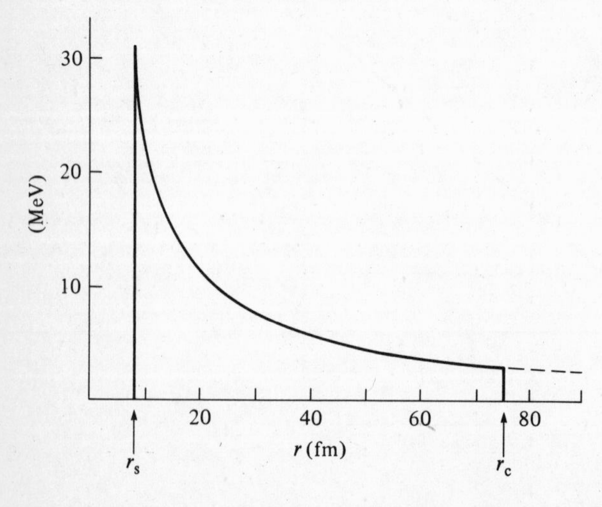

classical mechanics forbids the α-particle to escape. Quantum mechanics, however, allows it to tunnel through, and we now estimate this important tunnelling probability.

At distances $r > r_s$, outside the range of the strong interaction, the Schrödinger equation for the radial wave-function $u(r)$ of the α-particle is

$$-\frac{\hbar^2}{2m}\frac{1}{r}\frac{d^2}{dr^2}(ru)+\left[\frac{2Z_d e^2}{(4\pi\varepsilon_0)r}+\frac{\hbar^2}{2m}\frac{l(l+1)}{r^2}\right]u=Qu. \tag{6.3}$$

Here Z_d is the atomic number of the daughter nucleus, and m is the reduced mass, i.e.,

$$m=\frac{m_\alpha m_d}{m_\alpha+m_d}, \tag{6.4}$$

where m_α, m_d are the masses of the α-particle and daughter nucleus. The use of the reduced mass takes into account the recoil of the daughter nucleus. To conserve angular momentum, the angular momentum of the α-particle, and the angular momentum of the daughter nucleus, must combine to give the angular momentum of the parent. Also the parity of the final state, $(-1)^l \times$ (parity of daughter nucleus), must equal the parity of the parent nucleus. In the example $\mathrm{Bi}(\frac{9}{2}^-) \to \mathrm{Tl}(\frac{1}{2}^+)$, it is possible to conserve angular momentum with $l=4$ or $l=5$ (Appendix C), but parity conservation requires $l=5$.

For simplicity we shall only consider the case $l=0$. In fact the angular momentum term is usually small compared with the Coulomb potential. Writing $u(r)=f(r)/r$, equation (6.4) then reduces to

$$-\frac{\hbar^2}{2m}\frac{d^2 f}{dr^2}+\frac{2Z_d e^2}{(4\pi\varepsilon_0)r}f=Qf. \tag{6.5}$$

If the Coulomb term were replaced by a *constant* potential V_0 we should have solutions

$$f(r)=\begin{cases} e^{\pm ikr}, & Q>V_0,\quad k^2=\dfrac{2m}{\hbar^2}(Q-V_0) \\[2ex] e^{\pm Kr}, & Q<V_0,\quad K^2=\dfrac{2m}{\hbar^2}(V_0-Q). \end{cases} \tag{6.6}$$

This suggests that we try to find solutions to equation (6.5) of the form

$$f(r)=e^{\phi(r)}, \tag{6.7}$$

where $\phi(r)$ is to be determined. By substitution, $\phi(r)$ satisfies

$$\frac{\hbar^2}{2m}\left[\frac{d^2\phi}{dr^2}+\left(\frac{d\phi}{dr}\right)^2\right]=\left(\frac{2Z_d e^2}{(4\pi\varepsilon_0)r}-Q\right). \tag{6.8}$$

In a constant potential, $d^2\phi/dr^2 = 0$. Let us suppose that the Coulomb potential is sufficiently slowly varying for $d^2\phi/dr^2$ to be neglected in equation (6.8). Then

$$\frac{d\phi}{dr} = \pm\sqrt{\left[\frac{2m}{\hbar^2}\left(\frac{2Z_d e^2}{(4\pi\varepsilon_0)r} - Q\right)\right]}$$

and

$$\phi(r) = \pm\int\sqrt{\left[\frac{2m}{\hbar^2}\left(\frac{2Z_d e^2}{(4\pi\varepsilon_0)r} - Q\right)\right]}\, dr. \tag{6.9}$$

For $r > r_c$, we can write our approximate solution in the form

$$f(r) = A \exp\left(+i\int_{r_c}^{r} k(r)\, dr\right) + B \exp\left(-i\int_{r_c}^{r} k(r)\, dr\right), \tag{6.10}$$

where

$$k(r) = +\sqrt{\left[\frac{2m}{\hbar^2}\left(Q - \frac{2Z_d e^2}{(4\pi\varepsilon_0)r}\right)\right]},$$

and for $r_s < r < r_c$ we can write

$$f(r) = C \exp\left(+\int_{r}^{r_c} K(r)\, dr\right) + D \exp\left(-\int_{r}^{r_c} K(r)\, dr\right), \tag{6.11}$$

where

$$K(r) = +\sqrt{\left[\frac{2m}{\hbar^2}\left(\frac{2Z_d e^2}{(4\pi\varepsilon_0)r} - Q\right)\right]}.$$

A, B, C, D are constants to be determined by the boundary conditions.

The solution (6.10) represents outgoing and incoming waves. The incoming wave would be needed in an analysis of the scattering of α-particles by a nucleus. In the problem of α-decay, only the outgoing wave is present, so that $B = 0$.

This solution must then be matched on to the exponentially increasing and decreasing functions included in (6.11). Any admixture of the second term in the expression (6.11) quickly becomes negligible as r decreases, since the fall of the exponential is very rapid for typical values of the parameters, so that in the region $r_s < r < r_c$ we may take

$$f(r) = C \exp\left(\int_{r}^{r_c} K(r)\, dr\right). \tag{6.12}$$

At $r = r_s$, we must then match this solution to the appropriate radial function of the α-particle in the region where it is subject to the strong interaction. Here, at the surface of the daughter nucleus, the description of the α-particle is really a very complicated many-body problem. But it is reasonable to assume that in a heavy nucleus the rate of formation of α-

particles is a property of the nuclear surface and does not vary greatly from one nucleus to another. Given that an α-particle has been formed, the radial probability density of finding the particle at $r=r_c$ relative to the radial probability density of finding it at $r=r_s$ is given by

$$\frac{4\pi r_c^2|u(r_c)|^2}{4\pi r_s^2|u(r_s)|^2}=\frac{|f(r_c)|^2}{|f(r_s)|^2}. \tag{6.13}$$

We can therefore interpret

$$\left|\frac{f(r_c)}{f(r_s)}\right|^2=e^{-G}, \quad \text{say,} \tag{6.14}$$

as the transmission probability, through the Coulomb barrier, for α-particles created at $r\approx r_s$. The essential correctness of this interpretation is confirmed by a more exact analysis of the wave-functions and matching conditions.

Using equation (6.12)

$$G=2\int_{r_s}^{r_c} K(r)\,\mathrm{d}r$$

$$=2\int_{r_s}^{r_c}\sqrt{\left[\frac{2m}{\hbar^2}\left(\frac{2Z_d e^2}{(4\pi\varepsilon_0)r}-Q\right)\right]}\,\mathrm{d}r$$

$$=2\sqrt{\frac{2mQ}{\hbar^2}}\int_{r_s}^{r_c}\left(\frac{r_c}{r}-1\right)^{\frac{1}{2}}\mathrm{d}r,$$

since $r_c=\dfrac{2Z_d e^2}{(4\pi\varepsilon_0)Q}$.

With the substitution $r=r_c\cos^2\theta$ the integral is easily evaluated to give

$$G=2r_c\sqrt{\frac{2mQ}{\hbar^2}}\int_0^{\theta_0}2\sin^2\theta\,\mathrm{d}\theta$$

$$=2r_c\sqrt{\frac{2mQ}{\hbar^2}}\,[\theta_0-\sin\theta_0\cos\theta_0], \quad \text{where } \theta_0=\cos^{-1}\left[\sqrt{\frac{r_s}{r_c}}\right]$$

$$=\frac{\pi}{\hbar c}\left(\frac{2Z_d e^2}{4\pi\varepsilon_0}\right)\sqrt{\frac{2mc^2}{Q}}\,\mathscr{G}(r_s/r_c), \tag{6.15}$$

where the function

$$\mathscr{G}(r_s/r_c)=\frac{2}{\pi}\left[\cos^{-1}\left[\sqrt{\frac{r_s}{r_c}}\right]-\sqrt{\left[\frac{r_s}{r_c}\left(1-\frac{r_s}{r_c}\right)\right]}\right] \tag{6.16}$$

is dimensionless and lies between 1 and 0 for $0<r_s/r_c<1$ (Fig. 6.3). At low energies $r_c\to\infty$ and $\mathscr{G}\to 1$.

If the total flux of α-particles created at r_s is τ_0^{-1}, the probability per unit time of α-particle emission is $\tau_0^{-1}\, e^{-G}$, and hence the mean life for α-decay is given by

$$\tau = \tau_0\, e^G. \tag{6.17}$$

We have argued that the rate of formation of α-particles, and hence τ_0, is a nuclear property unlikely to vary greatly from nucleus to nucleus. In Table 6.1 we compare the formula (6.17) with experiment for a sequence of naturally occurring α-decays (all with $l=0$) initiated by the most common isotope of uranium $^{238}_{92}U$, taking τ_0 to be a constant. The value $\tau_0 = 7.0 \times 10^{-23}$ s was chosen to give a reasonable fit to this sequence of measured lives. This value is not unreasonable on a nuclear timescale of $2.6 \times 10^{-23} \times A^{\frac{1}{3}}$ s (§ 5.2), though it is somewhat shorter than earlier estimates because we have chosen r_s to be consistent with the modern values for nuclear radii. Early workers obtained estimates of nuclear radii from assumed values of τ_0.

The qualitative agreement of the simple theory (which was proposed in 1928 by Gamov and by Condon and Gurney) with experiment is truly remarkable. The simple quantum-mechanical formula for tunnelling comprehends timescales from as long as the age of the Earth (1.45×10^{17} s) down to times less than a microsecond. The largest discrepancy between theory and experiment occurs with $^{210}_{84}Po$; this discrepancy can be associated with the closed shell $N = 126$ in this isotope.

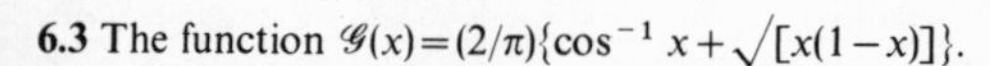
6.3 The function $\mathcal{G}(x) = (2/\pi)\{\cos^{-1} x + \sqrt{[x(1-x)]}\}$.

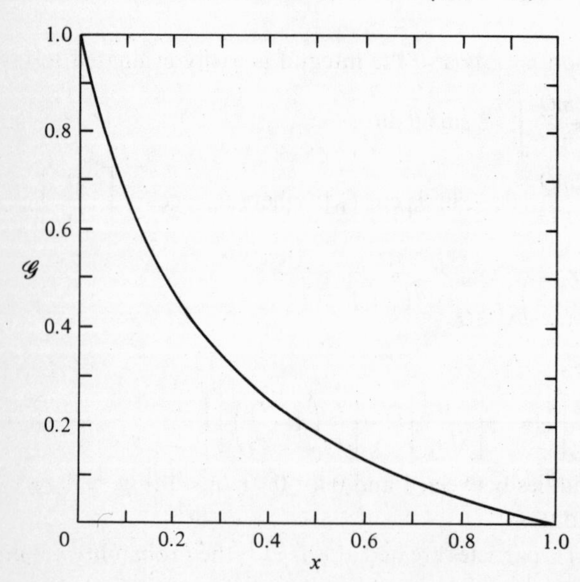

This particular sequence of decays is of interest in the early history of the study of radio-activity. As is indicated in Table 6.1, the daughter nucleus of an α-decay may be unstable to β-decay. Since the mass number A decreases by four in an α-decay, and is unchanged in β-decay, we expect there to be three other similar sequences of α-decays. Two of these, based on $^{232}_{90}$Th and $^{235}_{92}$U, are also naturally occurring. The third, initiated by $^{237}_{93}$Np, is made up of comparatively short-lived isotopes which must be produced artificially for the series to be observed.

Since mean lives are dominated by the tunnelling factor, and this in turn depends principally on the value of Q, we can now understand why decay rates with $Q<4$ MeV are so low.

It is found experimentally that α-particle emission can take place with the daughter nucleus left in an excited state. With even–even nuclei such processes usually occur with much lower probability, since the value of Q is reduced. However, the situation tends to be more complicated in the case of even–odd and odd–odd nuclei. An unpaired nucleon is less likely to take part in α-particle formation, and its state may form part of an excited configuration of the daughter nucleus. In such a case, the daughter nucleus is likely to be found in this excited state after the emission of an α-particle.

6.3 Spontaneous fission

In α-decay, a heavy nucleus splits into a light helium nucleus and another heavy nucleus. Fission is the name given to a similar but more

Table 6.1. *The α-decay series from $^{238}_{92}$U*

	Q (MeV)	r_s (fm)	r_c (fm)	$\mathscr{G}$	τ_{exp} (s)	τ_{theory} (s)
$^{238}_{92}U \rightarrow ^{234}_{90}Th$	4.27	8.52	60.7	0.53	2.0×10^{17}	3.3×10^{17}
		($^{234}_{90}Th \rightarrow ^{234}_{91}Pa \rightarrow ^{234}_{92}U$)				
$^{234}_{92}U \rightarrow ^{230}_{90}Th$	4.86	8.49	53.3	0.51	1.1×10^{13}	1.1×10^{13}
$^{230}_{90}Th \rightarrow ^{226}_{88}Ra$	4.77	8.45	53.1	0.51	3.5×10^{12}	3.9×10^{12}
$^{226}_{88}Ra \rightarrow ^{222}_{86}Rn$	4.87	8.41	50.9	0.50	7.4×10^{10}	7.4×10^{10}
$^{222}_{86}Rn \rightarrow ^{218}_{84}Po$	5.59	8.37	43.3	0.46	4.8×10^{5}	4.2×10^{5}
$^{218}_{84}Po \rightarrow ^{214}_{82}Pb$	6.11	8.33	38.7	0.43	2.6×10^{2}	1.6×10^{2}
		($^{214}_{82}Pb \rightarrow ^{214}_{83}Bi \rightarrow ^{214}_{84}Po$)				
$^{214}_{84}Po \rightarrow ^{210}_{82}Pb$	7.84	8.28	30.1	0.36	2.3×10^{-4}	1.1×10^{-4}
		($^{210}_{82}Pb \rightarrow ^{210}_{83}Bi \rightarrow ^{210}_{84}Po$)				
$^{210}_{84}Po \rightarrow ^{206}_{82}Pb$	5.41	8.24	43.7	0.47	1.7×10^{7}	5.8×10^{5}

The values of Q are from experiment. The intervening β-decays, which reduce the neutron-to-proton ratio as the nuclei become lighter, are given in parentheses.

symmetric process of a nucleus splitting into two more or less equal masses. The two pieces are called fission fragments.

The fragments are often nuclei in quite highly excited states, but we can estimate the energy release in fission by considering the simple case of the symmetric fission of an even–even nucleus (A, Z) into two identical nuclei $(A/2, Z/2)$ in their ground states, and using the semi-empirical mass formula (4.5). We shall neglect the pairing energies. For a fixed ratio of protons to neutrons, the symmetry term as well as the cohesive energy term is proportional to the total number of nucleons. Thus only the surface and Coulomb energies contribute to the difference in binding energy ΔB of the two fragments and the parent nucleus:

$$\begin{aligned}\Delta B &= 2B(A/2, Z/2) - B(A, Z)\\ &= -bA^{\frac{2}{3}}[2(\tfrac{1}{2})^{\frac{2}{3}} - 1] - \frac{dZ^2}{A^{\frac{1}{3}}}[2(\tfrac{1}{2})^{\frac{5}{3}} - 1].\end{aligned} \tag{6.18}$$

If ΔB is positive then this fission is energetically possible, and the fragments will acquire kinetic energy ΔB. From equation (6.18) nuclei for which

$$\frac{Z^2}{A} > \frac{b(2 - 2^{\frac{2}{3}})}{d(2^{\frac{2}{3}} - 1)}, \quad \text{i.e.,} \quad \frac{Z^2}{A} > 18, \tag{6.19}$$

are metastable with respect to fission. This condition is satisfied by β-stable nuclei heavier than ${}^{98}_{42}Mo$. The energy release on the fission of the heavy elements is much larger than that in α-decay. For example, the energy release in the symmetric fission of ${}^{238}_{92}U$ is $\simeq 180$ MeV. However, although this is large, the process is strongly inhibited by the tunnelling factor and spontaneous fission is only observed in the heaviest of elements.

We envisaged the process of α-decay as the initial formation of an α-particle at the surface of a nucleus and a subsequent tunnelling to freedom. It is not so easy to envisage the fission process, or to calculate the potential barrier. Figure 6.4 gives a schematic representation of the fission process, in which the nucleus is treated as a liquid drop. For the early stages of the fission of an initially spherical nucleus of radius R, it is reasonable to consider the nucleus deforming into an ellipsoid of revolution as in the first step of Fig. 6.4.

6.4 A schematic representation of a symmetric fission in the liquid drop model.

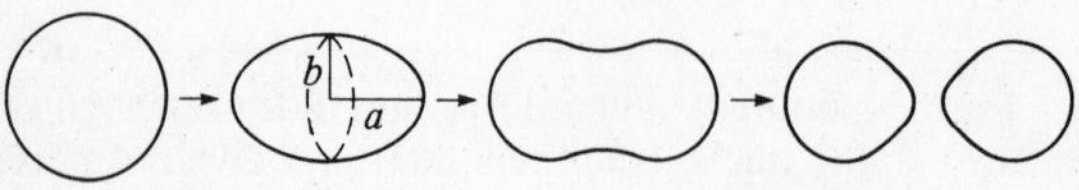

If we introduce a deformation parameter ε such that

$$\begin{aligned} a &= (1+\varepsilon)R \\ b &= R/(1+\varepsilon)^{\frac{1}{2}} \end{aligned} \tag{6.20}$$

where a, b are the major and minor semi-axes of the ellipsoid, the volume of the drop $(\frac{4}{3})\pi ab^2$ stays the same. It is not difficult to show that, for small ε, the surface area becomes to order ε^3

$$S(\varepsilon) = 4\pi R^2(1 + \tfrac{2}{5}\varepsilon^2 - \tfrac{52}{105}\varepsilon^3), \tag{6.21}$$

and the surface energy will, correspondingly, increase. (For small ε, the ε^3 term is a small correction.)

The Coulomb energy, on the other hand, decreases on deformation. A (rather lengthy) calculation for a uniformly charged ellipsoid gives the result

$$\begin{aligned} E_c &= \frac{\rho^2}{(4\pi\varepsilon_0)} \frac{1}{2} \iint \frac{d^3\mathbf{r}\, d^3\mathbf{r}'}{|\mathbf{r}-\mathbf{r}'|} \\ &= \frac{3}{5} \frac{(Ze)^2}{4\pi\varepsilon_0 R} (1 - \tfrac{1}{5}\varepsilon^2 + \tfrac{4}{21}\varepsilon^3) \end{aligned} \tag{6.22}$$

for small ε (cf. equation (4.6)).

Using the parameters of the semi-empirical mass formula, these results suggest that a small ellipsoidal deformation of a spherical nucleus gives a change in energy of

$$\varepsilon^2\left(\frac{2}{5}bA^{\frac{2}{3}} - \frac{1}{5}d\frac{Z^2}{A^{\frac{1}{3}}}\right) - \varepsilon^3\left(\frac{52}{105}bA^{\frac{2}{3}} - \frac{4}{21}d\frac{Z^2}{A^{\frac{1}{3}}}\right). \tag{6.23}$$

The coefficient of ε^2 is negative if

$$\frac{Z^2}{A} > \frac{2b}{d} = 51. \tag{6.24}$$

Hence when this condition is satisfied the deformation energy is negative even for small ε, and fission would proceed uninhibited by any potential barrier. Thus (6.24) suggests there is an absolute upper limit for chemical elements of $Z = 144$ (using the relation between Z and A for β-stable nuclei given by equation (4.9)).

For elements of lower Z, spontaneous fission involves tunnelling through a potential barrier. We can crudely estimate the height of the barrier from the expansion (6.23). For $^{235}_{92}U$ this gives the deformation energy

$$(83.35\varepsilon^2 - 159.16\varepsilon^3)\ \text{MeV}.$$

The coefficient of ε^3 is negative, which confirms that in the liquid drop model the most likely deformation is indeed a prolate, rather than oblate, ellipsoid, and the expression has a maximum of 3.4 MeV when $\varepsilon = 0.35$. The measured potential barrier is 5.8 MeV. For $A \sim 240$ barrier heights are found to be between 5 and 6 MeV. Experimental values are determined from the threshold energies required to induce fission, when the nucleus is bombarded with, for example, γ-rays. Induced fission by neutron capture also gives information on barrier heights. We shall consider induced fission in more detail in Chapter 8 and Chapter 9. It is a subject of great technological importance.

As the inequalities (6.19) and (6.24) indicate, Z^2/A is a measure of the likelihood that a nucleus will be subject to spontaneous fission. An empirical, approximately linear, relationship exists between the logarithm of the mean life for spontaneous fission and Z^2/A. This is shown in Fig. 6.5 for some even–even nuclei.

The fragments produced in spontaneous fission move apart rapidly

6.5 Mean lives for spontaneous fission of some even–even nuclei. (Data from *American Institute of Physics Handbook*, 3rd ed., 1972, New York: McGraw-Hill.)

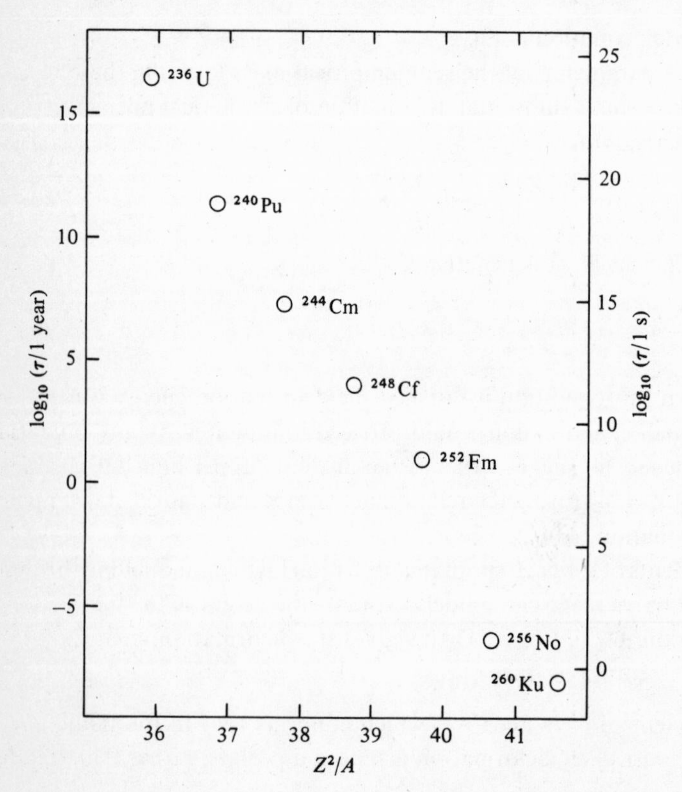

because of their Coulomb repulsion. They are neutron rich, since the equilibrium neutron-to-proton ratio of a β-stable nucleus decreases as A decreases, and they are in highly excited states. Typically, one to four neutrons 'boil off' from the fragments in a time of 10^{-18} to 10^{-15} s. Studies of the angular distribution of these 'prompt neutrons' show that they are indeed emitted from the moving fragments, rather than at the moment of break-up of the fissioning nucleus. The resulting nuclei are still far from the line of β-stability. They reach their ground states through the emission of prompt γ-rays and gradually decay, by β-emission, to stable nuclei. It occasionally happens that a nucleus produced by β-decay is unstable to neutron emission, and a 'delayed neutron' may result. For example, one of the fission products of ^{236}U is $^{87}_{35}$Br. This β-decays with a mean life of 80 s to either the ground state $^{87}_{36}$Kr, or an excited state $^{87}_{36}$Kr* which can lie above the threshold for neutron emission. In the latter case the rapid decay $^{87}_{36}\text{Kr}^* \to {}^{86}_{36}\text{Kr} + \text{n}$ sometimes occurs. Thus the timescale for the emission of the delayed neutrons is determined by the long lifetimes of the β-decay processes involved. We shall see in Chapter 8 that such processes are important for the control of nuclear reactors.

The semi-empirical mass formula predicts that the energy release in spontaneous fission is at a maximum when the two fragments are of equal mass. Notwithstanding this, it is usually found that there is a quite striking asymmetry in the mass distribution of the fission fragments. It is likely that this asymmetry is due to shell structure effects. More detailed theories of fission include these aspects of nuclear structure.

Problems

6.1 $^{8}_{4}$Be decays to two α-particles with a kinetic energy release of 0.094 MeV. Estimate its mean life using the tunnelling formula (6.17), and compare your estimate with the observed mean life of 2.6×10^{-17} s.

6.2 The isotope $^{194}_{79}$Au undergoes β-decay and has a mean life of 56 hours.

(*a*) One mode of decay is

$^{194}_{79}\text{Au} \to {}^{194}_{78}\text{Pt} + \text{e}^+ + \nu + 1.5\ \text{MeV}.$

The positron in this decay is created in the nucleus and must tunnel through a Coulomb barrier to escape. Show that the barrier factor suppresses the decay rate to a positron with an energy ~ 1 MeV only by a factor of about four.

(*b*) Another energetically possible decay, which has not been observed, is

$^{194}_{79}\text{Au} \to {}^{190}_{77}\text{Ir} + {}^{4}_{2}\text{He} + 1.8\ \text{MeV}.$

Estimate the mean life for this mode of decay.

6.3 ^{238}Pu decays by α-emission:

$$^{238}_{94}\text{Pu} \rightarrow ^{234}_{92}\text{U} + \alpha + 5.49 \text{ MeV},$$

with a mean life of 128 years. The mean life of ^{234}U is much longer, 2.5×10^5 years. Space probes to the outer planets use ^{238}Pu as a power source for their equipment. Estimate the mass of ^{238}Pu needed to supply a minimum of 1 kW of heat for 50 years.

6.4 The intermediate members of the radioactive series stemming from ^{238}U have negligible mean lives on geological time scales (Table 6.1), so that ^{238}U may be said to decay to ^{206}Pb with a mean life of 6.48×10^9 years. Similarly, ^{235}U decays to ^{207}Pb with a mean life of 1.03×10^9 years.

In a certain sample of uranium-bearing rock the proportions of atoms of ^{238}U, ^{235}U, ^{206}Pb, ^{207}Pb were measured to be 1000:7.19:79.7:4.85. The rock contained a negligible amount of ^{208}Pb, which is usually the most common isotope of lead, indicating that all the lead in the rock came from uranium decay. Estimate the age of the rock.

6.5 Estimate the energy release and the velocity of the fragments in the spontaneous fission

$$^{238}_{92}\text{U} \rightarrow ^{119}_{46}\text{Pd}^* + ^{119}_{46}\text{Pd}^*$$

where * denotes an excited state. (Use Fig. 4.7.)

Neutrons 'boil off' from the fragments. If in the frame of the moving fragments the neutrons are emitted isotropically with energy ≈ 2 MeV, describe qualitatively how the neutrons appear in the laboratory.

7

Excited states of nuclei

7.1 The experimental determination of excited states

So far we have for the most part been considering atomic nuclei in their quantum ground states. Most nuclei on Earth have been in their ground states since the time of its creation. However, almost all nuclei possess excited states of higher energy (and therefore less binding energy) than their ground state. There are many ways of exhibiting these excited states, and determining their energies and quantum numbers. One method is to scatter energetic protons of known momentum p_i from the nucleus of interest and to observe their angle of scattering θ and final momentum p_f. This process is illustrated in Fig. 7.1.

To conserve momentum, the recoiling nucleus has momentum $(p_i - p_f \cos\theta)$ in the direction of the incoming proton and $p_f \sin\theta$ in the perpendicular direction. Taking all momenta and energies to be non-relativistic, the difference E between the initial and final kinetic energies of the system is

$$E = \frac{p_i^2}{2m_p} - \frac{p_f^2}{2m_p} - \frac{(p_i^2 + p_f^2 - 2p_i p_f \cos\theta)}{2m_A^*}, \tag{7.1}$$

where m_A^* is the mass of the recoiling target nucleus. By conservation of energy E must be the excitation energy given to the nucleus. In terms of the initial and final proton kinetic energies E_i and E_f, equation (7.1) becomes

$$E = E_i\left(1 - \frac{m_p}{m_A^*}\right) - E_f\left(1 + \frac{m_p}{m_A^*}\right) + \frac{2m_p}{m_A^*}(E_i E_f)^{\frac{1}{2}} \cos\theta. \tag{7.2}$$

In equations (7.1) and (7.2), $m_A^* = m_A + E/c^2$ may be replaced by the mass m_A of the nucleus in its ground state, with little error.

In practice, a mono-energetic beam of protons is directed at a target containing the nucleus in question. If the target is a solid it is generally made so thin that the probability of a proton scattering more than once off a nucleus is small.

At a fixed scattering angle θ the emerging protons are no longer mono-energetic but, apart from a background coming from, for example, the residual multiple scattering, their energies fall into several well-defined peaks. An example of this is shown in Fig. 7.2.

In the experiment from which this data is taken, initial protons of energy 10.02 MeV were scattered from ${}^{10}_{5}B$, and the graph shows the number of protons scattered within a small angular range at $\theta = 90°$ as a function of their final energy E_f. The peak of the highest energy at $E_f = 8.19$ MeV corresponds to elastic scattering, since equation (7.2) then gives $E = 0$, that is, no excitation. The values of E_f for the successive peaks of lower energy give a sequence of excitation energies E of the ${}^{10}_{5}B$ nucleus (Problem 7.1).

7.1 Scattering of a proton from a nucleus initially at rest.

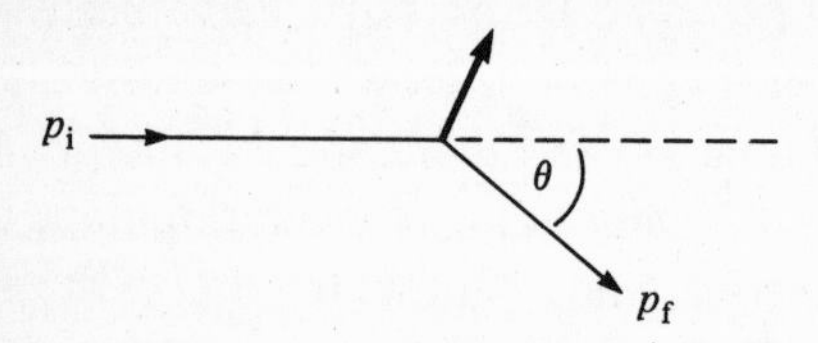

7.2 The number of protons scattered at 90° from a static target containing ${}^{10}B$, as a function of their final energy E_f. Initially the protons were in a collimated beam and had energy 10.02 MeV. Background scattering has been removed. (Data from Armitage, B. H. & Meads, R. E. (1962), *Nuc. Phys.* **33**, 494.)

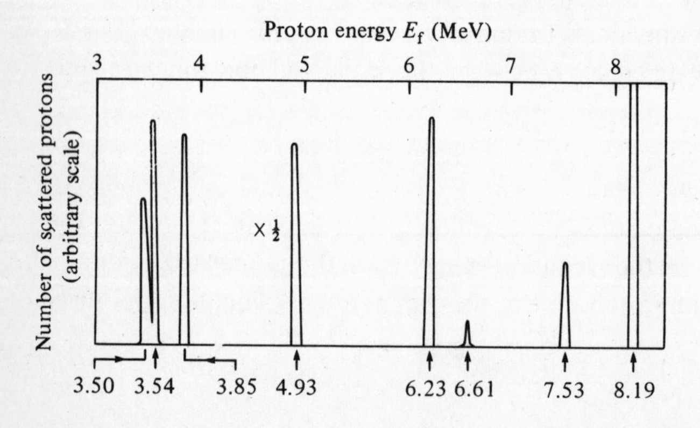

The area under the peak at a particular E_f in Fig. 7.2 is proportional to the probability of producing the corresponding excited state. This probability depends both on E_i and on θ. Information on the spin and parity of the state can be obtained from measurements of the angular dependence of the production probability. Further information on spin and parity is given by the energies and angular distribution of γ-rays that can result as the excited states decay back to the ground state.

The inelastic scattering of protons as in the above example is a technique which may be used with nuclei which are not radioactive and which can be safely made into targets. Another technique, which is suitable for determining the energy levels of some β-unstable nuclei also, is that of *deuteron stripping*.

In deuteron stripping, a mono-energetic beam of deuterons is directed at a target nucleus. As well as elastic and inelastic deuteron scattering, leaving the original, possibly excited, target nucleus, a *nuclear reaction* may take place in which the deuteron loses a nucleon into the target nucleus. Consider for example the reaction represented by

$$^{2}_{1}\mathrm{H} + {}^{A}_{Z}\mathrm{X} \rightarrow {}^{A+1}_{Z}\mathrm{X}^{*} + \mathrm{p}, \qquad (7.3)$$

in which only the proton emerges.

Here $^{A}_{Z}\mathrm{X}$ is the target nucleus and $^{A+1}_{Z}\mathrm{X}$ is its isotope (perhaps unstable) with one more neutron. The * denotes a possible excited state.

If the emerging proton in this reaction is at an angle θ with respect to the beam of incident deuterons and has energy E_f, a calculation similar to that for proton scattering yields for the excitation energy of the final nucleus, $^{A+1}_{Z}\mathrm{X}^{*}$, the expression

$$E = E_i\left(1 - \frac{m_d}{m^*_{A+1}}\right) - E_f\left(1 + \frac{m_p}{m^*_{A+1}}\right) + 2\,\frac{(m_p m_d E_i E_f)^{\frac{1}{2}}}{m^*_{A+1}} \cos\theta + E_0 \qquad (7.4)$$

where now E_i is the incident deuteron energy, m_d is the deuteron mass and $E_0 = (m_A + m_d - m_{A+1} - m_p)c^2$ is the difference in rest mass energies between the initial and final nuclei in their ground states.

Table 7.1 shows the results of a deuteron-stripping experiment,

$$^{2}_{1}\mathrm{H} + {}^{16}_{8}\mathrm{O} \rightarrow {}^{17}_{8}\mathrm{O}^{*} + \mathrm{p},$$

in which a deuteron beam with energy $E_i = 14.95$ MeV was directed at a target containing $^{16}_{8}\mathrm{O}$ and the energies of protons detected at $\theta = 19°$ were measured. In this example $E_0 = 1.93$ MeV. The table shows the six proton groups with the highest energies and the corresponding $^{17}_{8}\mathrm{O}$ excitation energies E. The highest-energy proton group with $E_f = 16.62$ MeV corresponds to the production of $^{17}\mathrm{O}$ in its ground state.

Figure 7.3 shows the excitation energies of ^{17}O up to 6 MeV. In this *energy-level diagram* the excited states are denoted by horizontal lines at a height above the ground state that is proportional to the excitation energy. The five lowest excited states are those determined from the above deuteron-stripping reaction. The experimental information on the others will be discussed in §8.1. We shall in general restrict our discussion to

Table 7.1

E_f (MeV)	11.42	11.97	12.69	13.50	15.74	16.62
E (MeV)	5.08	4.56	3.85	3.06	0.87	0.0

This shows the mean energies E_f of groups of protons that emerge, from a static target containing ^{16}O, at an angle of 19° to a 14.95 MeV deuteron beam. Below are the corresponding excitation energies E of ^{17}O, as calculated using equation (7.4). (Data from Yagi, K. *et al.* (1963), *Nuc. Phys.* **41**, 584.)

7.3 The ^{17}O energy level diagram up to an excitation energy of 5.94 MeV. The first five excited state energies are as determined from deuteron stripping (Table 7.1). Also shown is the threshold energy at 4.15 MeV for break-up into a neutron and ^{16}O (the 'neutron separation energy' of ^{17}O), and the threshold energy for break-up into ^{13}C and an α-particle. (For more information see Ajzenberg-Selove, F. (1982), *Nuc. Phys.* **A375**, 1.)

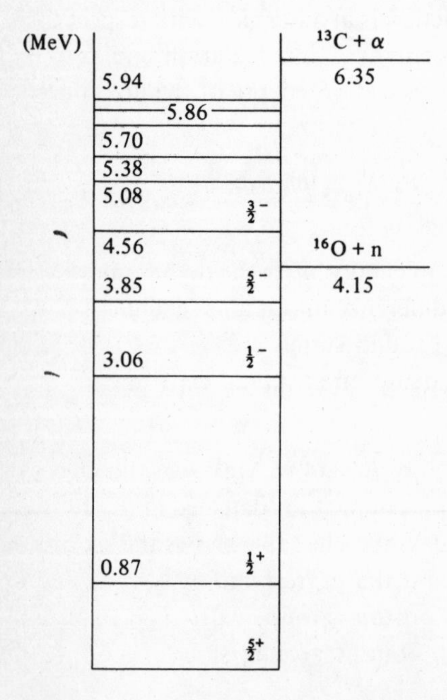

energy levels below about 10 MeV, which is the most important energy range for the topics we discuss in later chapters.

Also shown on the energy-level diagram are the lowest energies, called threshold energies, such that excited states above these thresholds can break up into the smaller nuclei indicated. These energies are computed from the masses of the nuclei involved. The lowest threshold is for ^{17}O to disintegrate into ^{16}O and a neutron. Below this threshold the excited states cannot disintegrate into lighter nuclei but they decay electromagnetically, for example by the emission of a photon, to lower energy states and, eventually, the ground state.

The spins and parities of the excited state, some of which are shown on the diagram, are deduced from measurements of the angular distribution of the protons from the nuclear reactions which produce the states, and also from the angular distributions of the photons resulting from the subsequent decays of the states.

7.2 Some general features of excited states

In general, the heavier the nucleus the more excited states it has. The deuteron has no excited states and very light nuclei have only a few well-defined excited states. However, the number of excited states increases rapidly as A increases. Figure 7.4 gives the energy levels up to 9 MeV of the two light nuclei $^{11}_{5}B$ and $^{11}_{6}C$. This pair is an example of so-called *mirror nuclei*: the number of protons in either one equals the number of neutrons in the other. The near equality of their energy levels illustrates the charge independence of the strong force; for this pair of light nuclei the difference in Coulomb energies is small and the nuclear physics is almost identical.

A qualitative understanding of the excited states is given by the shell model. Consider the $^{11}_{5}B$ nucleus. The six neutrons fill the $1s_{\frac{1}{2}}$ and $1p_{\frac{3}{2}}$ shells. There are two protons filling the $1s_{\frac{1}{2}}$ shell and in the $1p_{\frac{3}{2}}$ shell two protons have their angular momenta coupled to zero while the odd remaining proton gives the ground-state spin and parity $\frac{3}{2}^-$. The first excited state, spin and parity $\frac{1}{2}^-$, can be considered within the shell model to be the state in which the odd proton is taken from the $1p_{\frac{3}{2}}$ shell and placed in the higher-energy $1p_{\frac{1}{2}}$ shell. Such a state is known as a single-nucleon excitation.

Many of the higher energy states will correspond to several nucleon excitations. The fact that there is a large number of excited states is easily accommodated within the shell model. If we consider only the $1p_{\frac{3}{2}}$ and $1p_{\frac{1}{2}}$ shells, the four neutrons can be distributed in $\binom{6}{4}$ ways over the six available single-particle neutron states, and the three protons in $\binom{6}{3}$ ways over the single-particle proton states. Thus we can construct $\binom{6}{4} \times \binom{6}{3} = 15 \times 20 = 300$

independent states – more than enough to account for all of the states of negative parity below the α-decay threshold, even allowing for the fact that levels with spin j have $(2j+1)$ members.

Figure 7.5 shows energy-level diagrams for two heavier nuclei, ^{46}Ca and ^{108}Pd. Note in these examples that for a given excitation energy, the heavier nucleus has a greater density of excited states and, for a given nucleus, the density of states increases as the excitation energy increases. These qualitative features are apparent in most nuclei, though near to closed-shell nuclei the energy gaps between levels tend to be significantly greater, especially at low excitation energies. Again, the shell model provides an

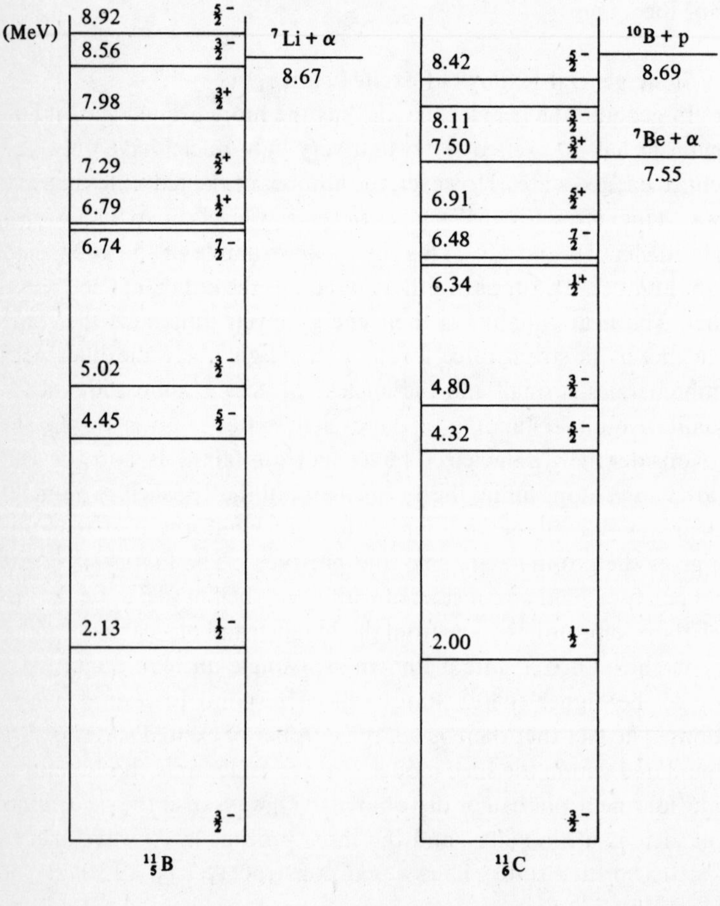

7.4 Energy-level diagrams for the mirror nuclei $^{11}_{5}B$ and $^{11}_{6}C$. The spins and parities of the states are also given. Note the proton separation energy from $^{11}_{6}C$ at 8.69 MeV, and the α-particle separation energies. (Data from Ajzenberg-Selove, F. & Busch, C. L. (1980), *Nuc. Phys.* **A336**, 1.)

(MeV)
8.92 $\frac{5}{2}^-$
8.56 $\frac{3}{2}^-$
$^7Li + \alpha$
8.67
7.98 $\frac{3}{2}^+$
7.29 $\frac{5}{2}^+$
6.79 $\frac{1}{2}^+$
6.74 $\frac{7}{2}^-$
5.02 $\frac{3}{2}^-$
4.45 $\frac{5}{2}^-$
2.13 $\frac{1}{2}^-$
$\frac{3}{2}^-$
$^{11}_{5}B$

8.42 $\frac{5}{2}^-$
$^{10}B + p$
8.69
8.11 $\frac{3}{2}^-$
7.50 $\frac{3}{2}^+$
$^7Be + \alpha$
7.55
6.91 $\frac{5}{2}^+$
6.48 $\frac{7}{2}^-$
6.34 $\frac{1}{2}^+$
4.80 $\frac{3}{2}^-$
4.32 $\frac{5}{2}^-$
2.00 $\frac{1}{2}^-$
$\frac{3}{2}^-$
$^{11}_{6}C$

explanation. The elementary formula (5.4) for the integrated density of single-nucleon states gives $\mathcal{N}(E) \sim E^{\frac{3}{2}}$ for neutrons or protons, so that the number of single-nucleon states $\Delta\mathcal{N}$ in a small energy range ΔE is given by

$$\frac{\Delta\mathcal{N}}{\Delta E} \approx \frac{\mathrm{d}\mathcal{N}}{\mathrm{d}E} = \frac{3}{2}\frac{\mathcal{N}}{E}. \tag{7.5}$$

Hence, taking $\Delta\mathcal{N} = 1$, the mean spacing between single-particle neutron levels at the Fermi energy ($E_{\mathrm{F}} \approx 38$ MeV, $\mathcal{N}(E_{\mathrm{F}}) = N$; see § 5.2) is

$$\Delta E = \frac{2}{3}\frac{E_{\mathrm{F}}}{N} \sim \frac{25}{N}\ \mathrm{MeV}, \tag{7.6}$$

with a similar result for the protons. ΔE very largely sets the energy scale for the excited states, so that as $N(\sim A/2)$ increases they come closer together.

In the shell model, the lowest-lying excitations can often be associated with single-particle excitations. At higher energies, several nucleons can be simultaneously excited, and the increasing density of states with energy reflects the increasing number of possible configurations involving many excited nucleons.

Often, such complex nuclear states can be quite simply described by models which naturally incorporate multi-particle motion. For example, the liquid drop with which we started our discussion of nuclei, and which we deformed in our discussion of spontaneous fission, can be envisaged to be in

7.5 Energy-level diagrams for ^{46}Ca and ^{108}Pd. (Data from *Nuclear Data Sheets* of the National Nuclear Data Centre for Nuclear Data Evaluation, Sheet 37 (1982), 290; Sheet 38 (1983), 467: Academic Press.)

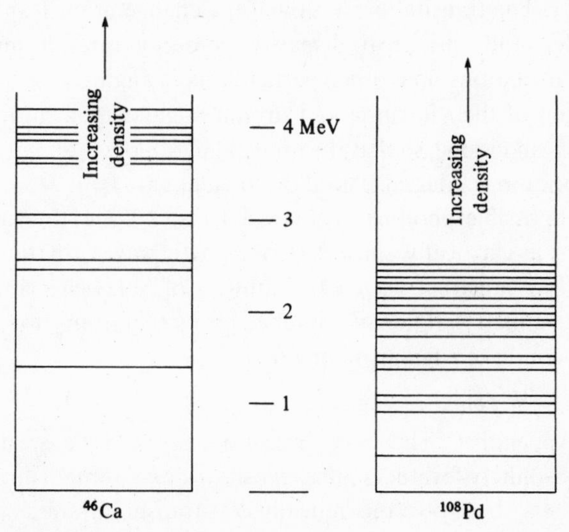

an excited state of vibration or one of overall rotation. Although we will not dwell here on these interesting and useful models, many excited states which it would be clumsy to describe in terms of the shell model can be justifiably envisaged as vibrational and/or rotational states.

The excited states of nuclei are not stable. Their energies, being of the order of MeV for light nuclei and keV for heavy nuclei, are so high they play an insignificant role in terrestrial thermodynamics. At temperatures accessible in laboratories they decay to states of lower energy and ultimately to the ground state. We now take up the question of their modes of instability and their mean lives.

7.3 The decay of excited states: γ-decay and internal conversion

Excited states that have energies below the lowest threshold for break-up into lighter nuclei decay almost exclusively electromagnetically. The most prominent mode is *γ-decay*, in which the nucleus changes to one of its lower energy states and simultaneously emits a single photon. A nucleus can also decay by *internal conversion*, which is a process whereby electromagnetic energy liberated by the nucleus is taken up by an atomic electron which is ejected. The energy of the emitted particle, be it photon or electron, is the energy lost by the nucleus, with corrections for small recoil effects and, in the case of internal conversion, the electron's atomic binding energy.

Electromagnetic mean lives can be as long as hundreds of years, or as short as 10^{-16} s. The transitions are slow if the change in nuclear spin is large. To understand this great disparity in decay rates it must be appreciated that photons, like other particles, have angular momentum, which is the sum of their intrinsic and orbital angular momentum. The intrinsic photon spin is one, so that the total angular momentum quantum number j of a photon is integral. The allowed values are $j = 1, 2, 3, \ldots$; the value $j = 0$ is not possible: photons do not exist in states of zero total angular momentum (just as classically, since electromagnetic waves are transverse, it is impossible to construct wave-like solutions of Maxwell's equations with spherical symmetry). If the nucleus changes its spin from j_i to j_f in a γ-decay, then to conserve angular momentum

$$j_i + j_f \geqslant j \geqslant |j_i - j_f|,$$

as is shown in Appendix C. Thus γ-ray transitions between states with $j_i = 0$ and $j_f = 0$ are absolutely forbidden (but transitions by internal conversion are possible). It may be shown theoretically that transition rates are much suppressed as j increases; the theory of γ-decay and internal conversion will be discussed more fully in Chapter 12.

As well as angular momentum, parity is conserved in electromagnetic transitions. The photon parity must be positive if the initial and final states have the same parity and negative if they have opposite parities. A photon has parity $(-1)^j$ when the decay is 'electric' with the nucleus basically coupling to the electric field of the photon, and parity $-(-1)^j$ when the decay is 'magnetic' with the nucleus coupling to the magnetic field of the photon.

Figure 7.6. shows the results of rough theoretical estimates of γ-decay rates. Precise calculations require a detailed knowledge of the initial and final nuclear wave functions, which is not generally available. As an example, consider the decay of the first excited state of $^{17}_{8}O$ (Fig. 7.3). This can only decay to the ground state and, neglecting internal conversion, will do so by emitting an 0.87 MeV photon. (See Problem 7.3 for recoil effects.) The nuclear spin changes from $\frac{1}{2}$ to $\frac{5}{2}$ and there is no change in nuclear

7.6 Estimated mean lives for electric multipole radiation of order 2^j as a function of the energy of the emitted photon, for a nucleus with $A = 100$. Corresponding estimates for other nuclei may be obtained by multiplying by $(100/A)^{2j/3}$. Mean lives for magnetic multipole radiation are generally longer than those for electric multipole radiation of the same order by a factor $\tau_M/\tau_E \sim 20A^{\frac{2}{3}}$.
(The lines are drawn from formulae given, for example, in Jackson, J. D. (1975), *Classical Electrodynamics*, 2nd ed., New York: Wiley, p. 760.)

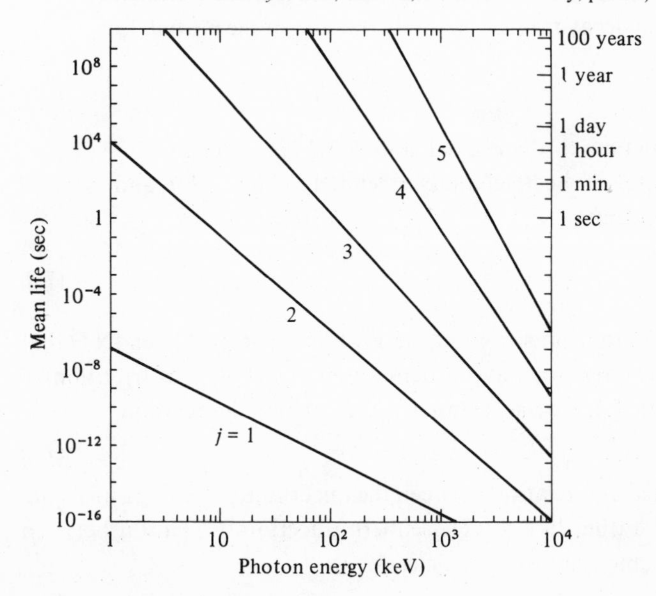

parity. Therefore the photon must have positive parity and $j \geqslant |\frac{1}{2}-\frac{5}{2}|=2$. The value $j=2$ is the most likely photon angular momentum; the value $j=3$ is possible but would give a much lower decay rate. The experimentally observed mean life is $(2.58 \pm 0.04) \times 10^{-10}$ s, in fair agreement with the value suggested by Fig. 7.6 for an electric transition with $j=2$.

Measurements of photon energies clearly give information on the energies of excited states, and such measurements have played a large part in determining these energies. Measurements of decay rates and of the angular distributions of the intensity and polarisation of the photons give information on the 'multipole' type of the transition. Transitions with $j=1, 2, 3, \ldots, n, \ldots$, are referred to as dipole, quadrupole, octapole, ..., 2^n-pole, ..., transitions; each type of transition has its characteristic lifetime and angular distribution. Unravelling the multipole type of a transition is one of the ways of determining the spins and parities of the nuclear states involved. Long-lived excited states of nuclei are known as *isomeric states*.

7.4 Partial decay rates and partial widths

In general, an excited state of a nucleus has the option of decaying in several ways. There may be several lower energy states to which it can decay by γ-emission, or it may be able to break up into lighter nuclei. For example, the 4.56 MeV excited state of ^{17}O (Fig. 7.3) can decay by neutron emission, or by γ-emission to any one of four lower energy levels. With each mode of decay, or *decay channel*, say the ith, there will be a partial decay rate $1/\tau_i$, and the total decay rate $1/\tau$ is simply the sum of the partial decay rates:

$$\frac{1}{\tau}=\sum_i \frac{1}{\tau_i}. \tag{7.7}$$

τ is the mean lifetime of the excited state (§ 2.3).

The *partial width* of the ith channel is defined to be $\Gamma_i=\hbar/\tau_i$ and the *total width* $\Gamma=\hbar/\tau$, so that

$$\Gamma=\sum_i \Gamma_i. \tag{7.8}$$

The Γ_i and Γ have the dimensions of energy. It is shown in Appendix D that an excited state does not have a definite energy, but a distribution of energies of width Γ about a mean energy E. Hence the relation

$$\Gamma\tau=\hbar \tag{7.9}$$

can be interpreted as a relation between the uncertainty in energy of a state and its lifetime, rather like the Heisenberg uncertainty relation between position and momentum of a particle.

The partial decay rates of nuclei for γ-emission are rarely greater than $10^{16}\ s^{-1}$. The corresponding partial widths are thus generally less than about 5 eV (and the energies of excited states that decay only by γ-emission, expressed in MeV, can be quoted to five decimal places).

Problems

7.1 From the data given in Fig. 7.2 draw an energy-level diagram for the nucleus ${}^{10}_{5}B$.

7.2 Derive equation (7.4).

7.3(*a*) Using the data of Table 7.1, show that the recoil velocity of a ${}^{17}O^*$ nucleus produced in its first excited level is

$v = 5.7 \times 10^{-3} c \quad (E_0 = 1.918 \text{ MeV}).$

(*b*) If this ${}^{17}O^*$ nucleus comes to rest before it decays, show that the energy of the emitted photon is about 24 eV less than the excitation energy of the nucleus.

(*c*) If the photon is emitted from the moving nucleus, show that because of the Doppler effect it will be changed in energy by between -5 keV and 5 keV.

7.4 The binding energies of the mirror nuclei ${}^{11}_{5}B$ and ${}^{11}_{6}C$ are 76.205 MeV and 73.443 MeV respectively. Assuming that the difference is due entirely to Coulomb effects, and that the proton charge is uniformly distributed through a sphere of radius R_C in both nuclei, find R_C. This was an early way of estimating the size of a nucleus. Compare R_C with the value $R = 1.1A^{\frac{1}{3}}$ fm, and comment on the difference.

7.5 The excited state ${}^{17}O^*$ at 4.56 MeV (Fig. 7.3) has a mean life of only 1.6×10^{-20} s. How can this be so short? Estimate the width Γ of the state.

7.6 What type of electromagnetic transition do you expect between a state at 2.13 MeV in ${}^{11}_{5}B$ (Fig. 7.4) and the ground state? Estimate the mean life of this state.

7.7 Consider the energy levels of ${}^{10}_{5}B$ (Problem 7.1). The ground state has spin and parity 3^+, and the excited states in order of increasing excitation energy are $1^+, 0^+, 1^+, 2^+, 3^+, 2^-, 2^+, \ldots$. Is there an explanation within the shell model of why the lowest states all have positive parity?

The first excited energy level is at 0.72 MeV and the second at 1.74 MeV. Given a large number of nuclei in the second excited state, what energies have the γ-rays that result from the decays? Estimate the relative numbers of these γ-rays.

8
Nuclear reactions

In a nuclear reaction two nuclei, or a nucleon and a nucleus, come together in such close contact that they interact through the strong force. The deuteron-stripping reaction of equation (7.3) is one example. A reaction which contributes to energy generation in stars is

$$^{16}O + {}^{16}O \rightarrow {}^{28}Si + \alpha + 9.6 \text{ MeV}, \tag{8.1}$$

and a nuclear reaction important in power technology is

$$n + {}^{235}U \rightarrow \text{fission products.}$$

The latter two are both *exothermic* reactions in which the kinetic energy of the final nuclei is greater than that of the initial nuclei. In an *endothermic* reaction energy must be supplied before the reaction will take place, as in the reaction inverse to (8.1) above.

8.1 The Breit–Wigner formula

The concept of *cross-section* (Appendix A) is important for understanding and classifying nuclear reactions. Figure 8.1 shows the total cross-section for neutrons to interact with the $^{16}_{8}O$ nucleus as a function of the kinetic energy E (in the centre-of-mass system) up to $E = 2.3$ MeV. The principal features of the cross-section are the high but narrow *resonance peaks*, superposed on a slowly varying background. These peaks are due to the formation of excited states of ^{17}O from the neutron and ^{16}O at the resonance energies. When the energy of the incident neutron is such that the total energy of the system matches, to within the width Γ, one of the excited states energies of ^{17}O, the neutron is readily accepted into the target to form that state. Note that the binding energy of the neutron in the ground state of the so-called *compound nucleus* becomes available as excitation energy. In our example of ^{17}O, if excitation energies are measured from the ground state the neutron binding energy of 4.15 MeV (cf. Table 4.2 and Fig. 7.3) has to be added to the resonance energies to obtain the ^{17}O excitation energies. This displaced energy scale is also given in Fig. 8.1. Thus only those excited states above 4.15 MeV can appear in the data.

The six peaks which appear in Fig. 8.1 correspond to the top six levels of the energy-level diagram, Fig. 7.3. The two lowest of these six correspond to states found in the deuteron-stripping reaction we discussed earlier in § 7.1.

It is shown in Appendix D that excited states make a contribution to the total cross-section in the neighbourhood of the resonance energy E_0 of approximately the form

$$\sigma_{\text{tot}}(E) = \frac{\pi}{k^2} \frac{g\Gamma_i\Gamma}{(E-E_0)^2 + \Gamma^2/4}, \tag{8.2}$$

where $k = |\mathbf{k}|$, and $\mathbf{k}$ is the wave-vector of the incoming neutron in the centre-of-mass frame, Γ_i is the partial width for decay into the incident

8.1 The total cross-section for neutrons interacting with ^{16}O as a function of centre-of-mass energy, showing resonances that correspond to the formation of excited states of ^{17}O (top scale: see also Fig. 7.3). (Data from Garber, D. I. & Kinsey, R. R. (1976), *Neutron Cross Sections*, vol. II, Upton, New York: Brookhaven National Laboratory.)

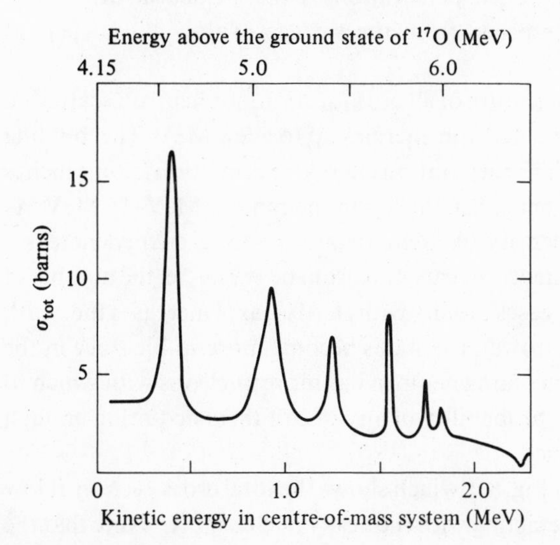

channel $^{16}O+n$, g is a statistical factor (in this case $g=(2j+1)/2$ where j is the spin of the excited state). The expression (8.2) is known as the *Breit–Wigner formula*. For $\Gamma \ll E_0$ the cross-section is at a maximum when $E=E_0$, and falls to half its maximum value at $E=E_0 \pm \Gamma/2$. Thus Γ is the 'full width at half-maximum' of the peak. The peak width seen experimentally depends also on the energy spread of the incident neutron beam (no particle beam is ever perfectly mono-energetic), on the thermal motion of the nuclei in the target, and on the characteristics of the detectors, so that a careful analysis may be necessary before a true intrinsic width can be obtained from the raw experimental data.

Consider the peak at $E=0.41$ MeV in Fig. 8.1. The estimated width of the peak, $\Gamma \approx 0.04$ MeV, corresponds to a mean life of $\approx 1.6 \times 10^{-20}$ s. This is short compared with mean lives for γ-emission, but still quite long on the nuclear time scale (§ 5.2) of the oxygen nucleus of $\sim 10^{-22}$ s. Such a long mean life can be understood as resulting from the nature of the excited state, which is a compound nuclear state in which many nucleons participate. The neutron entering the nucleus loses its energy by collisions with other nucleons, and if it loses more than 0.41 MeV it can no longer escape. The nucleus then stays in the excited state until such time as a single neutron again acquires enough energy to get away, or (in this case with much lower probability) the nucleus decays electromagnetically. In the latter case, if the decay is to the ground state or any other state below the neutron separation energy, the neutron is captured; this process is known as *radiative capture*.

In between resonance peaks, an analysis of the background cross-section suggests that the nucleus resists penetration by the incident neutron. The neutron appears to be repelled from the surface of the ^{16}O nucleus at energies off resonance.

Resonance peaks are a feature of all cross-sections for neutron scattering from nuclei with $A \geqslant 4$ and neutron energies up to a few MeV. The 'binding energy of the last neutron' (or separation energy: see equation (5.6)) which is available for excitation energy usually lies in the range 5 MeV–15 MeV. As explained in § 7.2, the density of excited states at fixed energy increases rapidly with A. Thus for neutron scattering from heavy nuclei the number of resonances per MeV increases rapidly with A. Also, as A increases the width of the states becomes narrower: the states become more stable since in the compound nucleus the incoming neutron has more nucleons with which to share its energy, and the probability of any one of them acquiring enough energy to escape decreases.

All this is illustrated in Fig. 8.2, which shows the total cross-section at low energies for neutrons interacting with the heavy nucleus $^{238}_{92}U$. Note that the horizontal energy scale is in electron volts, and the vertical cross-section scale is logarithmic. The resonance peaks are associated with the formation of excited states of ^{239}U, and the spacings between the peaks are only ≈ 20 eV. The resonances are very narrow, with an intrinsic width of order 10^{-2} eV. Indeed, the states are here so narrow that γ-decay competes significantly with other decay modes, and roughly half of the decays of the excited states formed at these resonances are electromagnetic and result in radiative capture. The other prominent decay mode is neutron emission. Less-common modes include α-decay and fission.

For neutron energies that are off resonance the cross-section of Fig. 8.2 is dominated by the neutron scattering from the surface of the ^{238}U nucleus. However, other nuclear reactions are energetically possible and may occur. For example, the neutron could pick up two protons and another neutron from the ^{238}U surface to form an α-particle:

$$n + {}^{238}_{92}U \rightarrow {}^{4}_{2}He + {}^{235}_{90}Th.$$

Such a reaction, when the neutron energy is off resonance, does not proceed through the formation of ${}^{239}_{92}U^*$, and is known as a *direct nuclear reaction.*

8.2 Neutron reactions at low energies

Since neutrons are uncharged there is no Coulomb barrier to overcome; hence neutrons of very low energy easily penetrate matter and interact with nuclei. In the limit $E \rightarrow 0$, only elastic scattering and exothermic nuclear reactions can take place. When a nuclear reaction is possible it can be expected that the reaction rate at sufficiently low energies

8.2 The total cross-section for neutrons interacting with ^{238}U, as a function of centre-of-mass energy. Note that the vertical scale for the cross-section is logarithmic and the horizontal energy scale is in electron volts. (Data as in Fig. 8.1.)

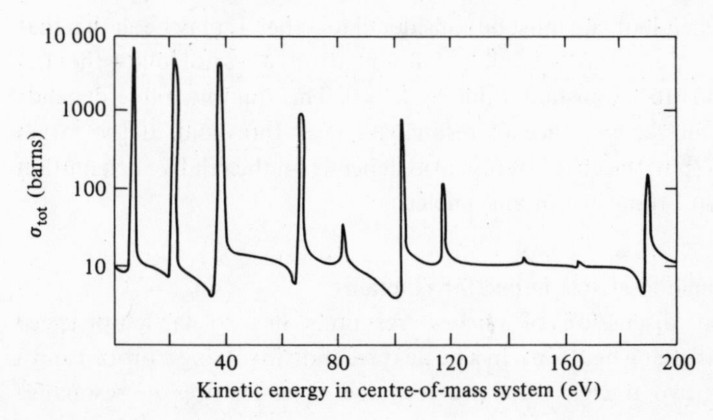

will be independent of E, and simply proportional to the density of neutrons in the neighbourhood of the nucleus. The cross-section σ_{ex} for exothermic nuclear reactions is given (Appendix A) by

$$(\text{neutron flux}) \times \sigma_{ex} = \text{reaction rate per nucleus.}$$

The neutron flux is $\rho_n v$ (where ρ_n is the neutron number density in the beam and v is the velocity of the neutrons relative to the target nucleus). Since the right-hand side of the equation is also proportional to ρ_n, it follows that

$$\sigma_{ex} = \frac{(\text{constant})}{v} \tag{8.3}$$

at sufficiently low energies. This is the behaviour that is seen experimentally.

If the low-energy region lies in the wing of a resonance close to threshold, the $(1/v)$ law is consistent with the Breit–Wigner formula (8.2). However, in this case we must take into account the energy dependence of the partial width $\Gamma_i(E)$, found in equation (D.6) of Appendix D. $\Gamma_i(E)$ contains the factor $n_i(E)$, which is proportional to k. For $E \approx 0$, the Breit–Wigner formula then gives

$$\sigma \approx \frac{1}{k} \frac{(\text{constant})}{E_0^2 + \Gamma^2/4}. \tag{8.4}$$

Since $\hbar k = mv$, where m is the reduced mass of the neutron and target nucleus, we recover the $(1/v)$ law. The combination of the $(1/v)$, or, equivalently, $(1/E^{\frac{1}{2}})$ law with a low-lying resonance is well illustrated in Fig. 8.3, which shows the low-energy cross-section for cadmium.

In quantum mechanics, non-elastic processes are always accompanied by elastic scattering, just as, in optics, absorption is always accompanied by diffraction. The elastic scattering of neutrons by nuclei takes place through compound nucleus formation and by surface scattering; the two processes are not independent and must be considered together. It may be shown that the elastic-scattering cross-section of slow neutrons does not follow the $(1/v)$ law but tends to a constant value as $E \to 0$. This limiting value depends sensitively on the presence of resonances near threshold. If the target nucleus has spin, the cross-section also depends on the relative orientation of the spins of the neutron and nucleus.

8.3 Coulomb effects in nuclear reactions

Our discussion of nuclear reactions has so far emphasized reactions involving neutrons. In a nuclear reaction involving a proton and a nucleus, or two nuclei, there are seen the same features of resonance scattering with the formation of a compound nuclear state, and direct nuclear reactions off resonance.

However, the effect of the Coulomb repulsion between particles in the initial or final channels of the reaction leads to significant differences in the reaction cross-sections at low energies below the Coulomb barrier height. This effect is illustrated in Fig. 8.4, which shows the low-energy cross-section for the nuclear reaction

$$\alpha + {}^{13}_{6}\mathrm{C} \rightarrow \mathrm{n} + {}^{16}_{8}\mathrm{O},$$

as a function of the centre-of-mass kinetic energy E of the incident nuclei. The reaction is in fact exothermic, with an energy release of 2.2 MeV (Fig. 7.3) and so it can in principle occur at any energy. However, at low energies the α-particle must tunnel through a Coulomb barrier before it can interact with the ^{13}C nucleus. The barrier is about 4 MeV high. In classical mechanics a nuclear interaction could not occur for an α-particle having lower energy than this. In quantum mechanics the particle can tunnel

8.3 The total cross-section for neutrons interacting with natural cadmium. The open circles o are experimental points (data as in Fig. 8.1). The line is a fit with

$$\sigma_{\mathrm{tot}} = \frac{(\mathrm{constant})}{v[(E-E_0)^2 + \Gamma^2/4]},$$

taking $E_0 = 0.18$ eV and $\Gamma = 0.12$ eV. Note that both the scales are logarithmic. On a logarithmic plot the '$1/v$' form at very low energies gives a straight line with a slope of $-\frac{1}{2}$ (cf. equation (8.4), $v = (2mE)^{\frac{1}{2}}$). This is evident for $E < 0.03$ eV. The very large resonance cross-section is due to ^{113}Cd, which constitutes 12.3% of natural cadmium.

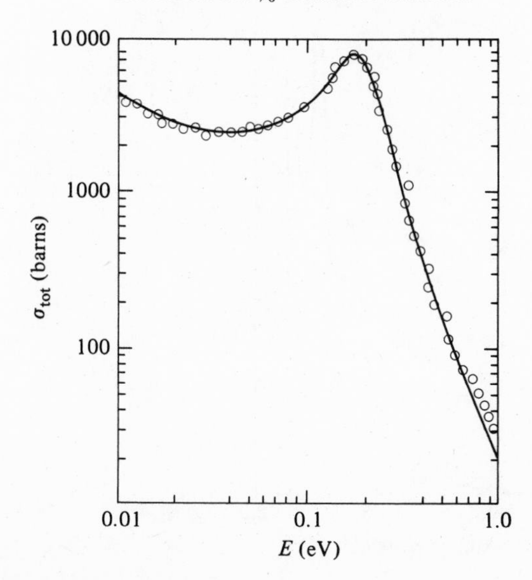

through, but the low-energy cross-section is much suppressed, as the figure clearly demonstrates. The tunnelling probability for the α-particle to penetrate the barrier from the outside is the same as the probability for tunnelling in the other direction, as in α-decay, and this we estimated in Chapter 6 to be $e^{-G(E)}$, where $G(E)$ is given by equation (6.15) (but with Q replaced by E).

It is usual to parametrise charged-particle reaction cross-sections at low energies by the expression

$$\sigma(E)=\frac{1}{E}S(E)e^{-G(E)}, \tag{8.5}$$

and Fig. 8.4 also shows this curve with $S(E)$ chosen to be a constant 0.3 barn MeV to fit the cross-section at the lowest energies. The background cross-section below the resonances roughly follows this curve, but large resonance peaks due to the formation of excited states of ^{17}O are evident.

The precise form of charged-particle nuclear reaction cross-sections at low energies is of great importance, both in astrophysics and for the

8.4 The cross-section for the reaction $\alpha+{}^{13}C\rightarrow n+{}^{16}O$. The dashed curve --- exhibits the large Coulomb suppression at low energies (see text). Note the resonances at high excited-state energies of ^{17}O (top scale) which are above those shown in Fig. 7.3. (Data from Blair, J. K. & Haas, F. X. (1973), *Phys. Rev.* **C7**, 1356.)

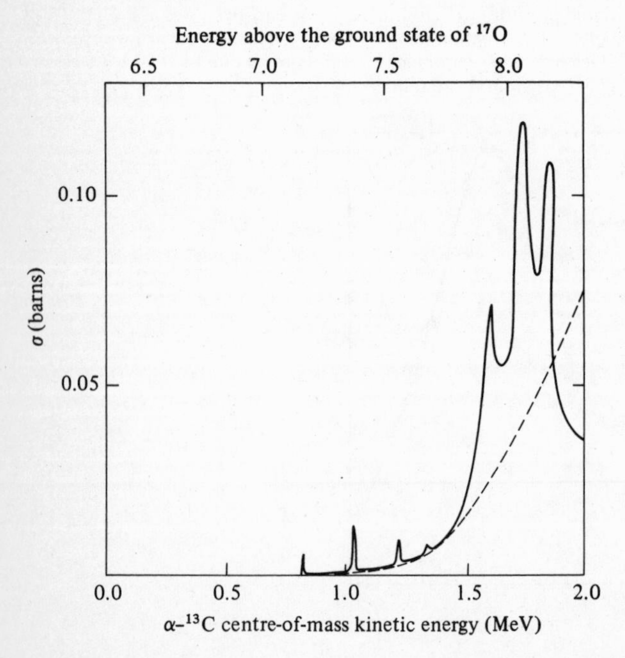

prospect of controlled thermonuclear reactions on Earth. It may be shown that, as $E \to 0$, the function $S(E)$ in (8.5) tends to a constant value, which depends on the particular reaction and is very sensitive to the proximity of resonances. We can give a qualitative derivation of this result for the case when the low-energy region lies in the wing of a nearby resonance. The Breit–Wigner formula (8.2) may be written

$$\sigma(E)=\frac{\pi\hbar^3}{2mE}\frac{g\Gamma(\Gamma_i/\hbar)}{(E-E_0)^2+\Gamma^2/4}, \tag{8.6}$$

where m is the reduced mass of the interacting particles and $\Gamma_i/\hbar$ is the decay rate into the incident channel. For energies close to threshold it is again important to include the energy dependence of Γ_i. Recalling the discussion of α-decay in Chapter 6, we replace the decay rate $(\Gamma_i/\hbar)$ by $(1/\tau_0)e^{-G(E)}$, where τ_0 is a constant nuclear time. For $E \approx 0$ the expression (8.6) then reduces to

$$\sigma(E)=\frac{\pi\hbar^3}{2m\tau_0}\frac{g\Gamma}{E_0^2+\Gamma^2/4}\frac{e^{-G(E)}}{E}.$$

This is of the same form as (8.5) with

$$S(E)=\frac{\pi}{2m}\frac{\hbar^3}{\tau_0}\frac{g\Gamma}{E_0^2+\Gamma^2/4},$$

a constant.

If the nuclei in a reaction have charges Z_1e, Z_2e, the expression (6.15) for $G(E)$ must of course be generalised slightly: $2Z_de^2$ is replaced by $Z_1Z_2e^2$, and m becomes the reduced mass of the nuclei involved. At very low energies r_c is large and so $\mathcal{G}(r_s/r_c) \to 1$. Thus

$$G(E)\approx\frac{\pi}{\hbar c}\left(\frac{Z_1Z_2e^2}{4\pi\varepsilon_0}\right)\sqrt{\frac{2mc^2}{E}}$$

$$=\sqrt{\frac{E_G}{E}}, \quad \text{say,}$$

where

$$E_G=2mc^2\left(\frac{\pi Z_1Z_2e^2}{\hbar c(4\pi\varepsilon_0)}\right)^2$$

and

$$\sigma(E)\approx\frac{1}{E}S(0)e^{-\sqrt{(E_G/E)}}. \tag{8.7}$$

8.4 Doppler broadening of resonance peaks

We mentioned in § 8.1 that the thermal motion of the nuclei in the target affects the width of a resonance as seen experimentally. Neutrons in a beam incident on a target, and mono-energetic with respect to that target, are not mono-energetic with respect to the individual nuclei in the target, since these will be in random thermal motion. The energy that appears in the Breit–Wigner formula is the energy in the centre-of-mass frame of the neutron and the target nucleus. If the neutron has velocity $\mathbf{v}_1$ and the nucleus velocity $\mathbf{v}_2$, the centre of mass energy is

$$E=\tfrac{1}{2}m(\mathbf{v}_1-\mathbf{v}_2)^2=E_1+\frac{m}{M}E_2-2\sqrt{\left(\frac{m}{M}E_1E_2\right)}\cos\theta,$$

where M is the mass of the nucleus, $m=m_nM/(m_n+M)$ is the reduced mass, E_1 is the centre-of-mass energy when thermal energy is neglected, E_2 is the thermal energy of the nucleus and θ the angle between $\mathbf{v}_1$ and $\mathbf{v}_2$.

The term $(m/M)E_2$ can be neglected if the neutron energy is much greater than thermal or if the target nucleus is heavy. The thermal energy E_2 is of order of magnitude k_BT, where k_B is Boltzmann's constant and T is the temperature of the target. Since $\cos\theta$ lies between -1 and $+1$, it can be seen that, when averaged over many nuclei, E will have a spread in energy about E_1 of magnitude

$$\Delta E\approx 2\sqrt{\left(\frac{m}{M}E_1k_BT\right)}.$$

Thus if a cross-section is measured in the laboratory as a function of E_1, in the neighbourhood of a resonance at energy E_0, the Breit–Wigner form is

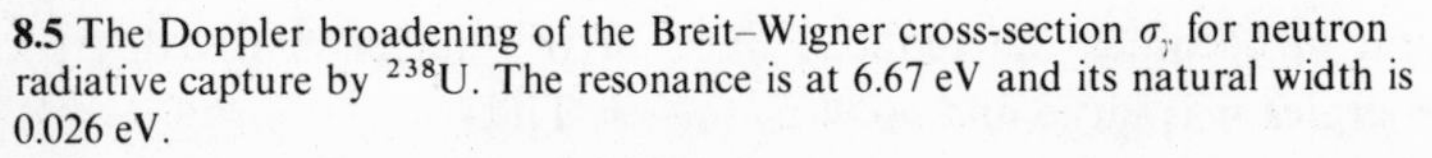

8.5 The Doppler broadening of the Breit–Wigner cross-section σ_γ for neutron radiative capture by ^{238}U. The resonance is at 6.67 eV and its natural width is 0.026 eV.

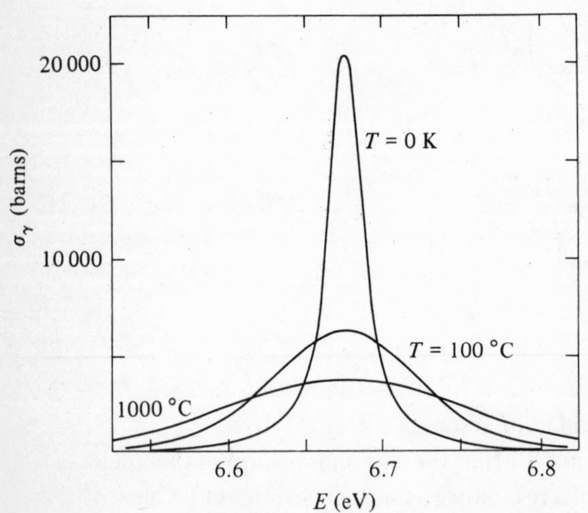

modified and, in particular, the width of the resonance peak will be larger than the natural width by an amount of order

$$\Delta\Gamma \approx 2\sqrt{\left(\frac{m}{M}E_0 k_B T\right)}.$$

This is *Doppler broadening*. A more detailed analysis shows that the total area under the resonance peak is independent of temperature, so that the height of the peak is reduced as the width increases. This is illustrated in Fig. 8.5 for a resonance in ^{238}U. (We shall see in Chapter 9 that Doppler broadening is of crucial importance for the thermal stability of nuclear reactors.) In the resonance peak of Fig. 8.3, on the other hand, it is easy to check that the effect of Doppler broadening at room temperature T_r ($k_B T_r \approx (1/40)$ eV) is small.

Problems

8.1 Quantum mechanics gives the total cross-section for scattering from an impenetrable sphere of radius R at low energies ($kR \ll 1$) to be $\sigma = 4\pi R^2$. For the cross-section of Fig. 8.2, show that the order of magnitude of the cross-section between resonances is given by this formula with R the radius of the uranium nucleus, and at a resonance the order of magnitude is given by $(\lambda/2\pi)^2$, where λ is the neutron wavelength ($\lambda = 2\pi/k$), as is implied by the Breit–Wigner formula (8.2).

8.2 Neutron detectors register individual neutrons by their production of charged, ionising particles in a nuclear reaction. One method, appropriate to thermal neutrons ($E < 0.1$ eV) uses the reaction

$$n + {}^3_2\text{He} \rightarrow p + {}^3_1\text{H} + 0.73 \text{ MeV}.$$

The cross-section for this reaction, which dominates at low energies, follows the $(1/v)$ law,

$$\sigma = 0.039(c/v) \text{ b}.$$

The mean distance a neutron travels through ^{3}He gas before it interacts is $l = 1/(\rho_{He}\sigma)$, where ρ_{He} is the number density of helium atoms (Appendix A). What detector thickness is needed, using ^{3}He gas at a pressure of 10 bars (which gives $\rho_{He} = 2.4 \times 10^{26}$ m^{-3}) in order that at least 90% of incident neutrons with energy 0.1 eV produce ionisation?

8.3 The nucleus $^5_3$Li is apparent as a resonance in the elastic scattering of protons from $^4_2$He at a proton energy ≈ 2 MeV. The resonance has a width of 0.5 MeV and spin $\frac{3}{2}$.

(*a*) What is the lifetime of $^5_3$Li?

(*b*) Estimate the cross-section at the resonance energy.

8.4 Figure 9.1(*b*) shows the measured total cross-section for neutrons incident on ^{238}U. What conclusion can you draw from the apparent absence of $(1/v)$ behaviour at low neutron energies?

8.5 The zero-temperature radiative capture cross-section illustrated in Fig. 8.5 is the intrinsic cross-section to which the Breit–Wigner formula is immediately applicable. The excited state has spin $\frac{1}{2}$, and there are two significant decay channels; the dominant one is γ-emission and the other is neutron emission. Estimate the relative probability of neutron radiative capture at resonance, and estimate the elastic neutron scattering cross-section at resonance. (Hint: use equation (D.11).)

9
Power from nuclear fission

We saw in Chapter 4 that nuclei in the neighbourhood of ^{56}Fe are the most strongly bound (Fig. 4.7). In principle therefore, nuclear potential energy can be released into kinetic energy and made available as heat by forming nuclei closer in mass to iron, either from heavy nuclei by fission or from light nuclei by fusion. This chapter is devoted to the physics of nuclear fission and its application in power reactors. In 1984, about 6% of the total energy consumed in the UK came from nuclear fission.

9.1 Induced fission

The spontaneous fission of nuclei such as ^{236}U was discussed in § 6.3; the Coulomb barriers inhibiting spontaneous fission are in the range 5–6 MeV for nuclei with $A \approx 240$. If a neutron of zero kinetic energy enters a nucleus to form a compound nucleus, the compound nucleus will have an excitation energy above its ground state equal to the neutron's binding energy in that ground state. For example, a zero-energy neutron entering ^{235}U forms a state of ^{236}U with an excitation energy of 6.46 MeV. This energy is clearly above the fission barrier, since the compound nucleus quickly undergoes fission, with fission products similar to those found in the spontaneous fission of ^{236}U. To induce fission in ^{238}U, on the other hand, requires a neutron with a kinetic energy in excess of about 1.4 MeV. The 'binding energy of the last neutron' in the nucleus ^{239}U is only 4.78 MeV, and an excitation energy of this amount clearly lies below the fission threshold of ^{239}U. The differences in the binding energy of the last neutron in even-A and odd-A nuclei are incorporated in the semi-empirical mass formula in the pairing energy term and are clearly evident in induced fission. The odd-A nuclei

$$^{233}_{92}U, \quad ^{235}_{92}U, \quad ^{239}_{94}Pu, \quad ^{241}_{94}Pu,$$

are examples of 'fissile' nuclei, i.e. nuclei whose fission is induced even by a zero energy neutron, whereas the even-A nuclei

$$^{232}_{90}Th, \quad ^{238}_{92}U, \quad ^{240}_{94}Pu, \quad ^{242}_{94}Pu$$

require an energetic neutron to induce fission.

9.2 Neutron cross-sections for ^{235}U and ^{238}U

The principal isotopes of naturally-occurring uranium are ^{235}U (0.72%) and ^{238}U (99.27%). Figure 9.1 shows the total cross-sections σ_{tot} and fission cross-sections σ_f of ^{235}U and ^{238}U for incident neutrons of energy E from 0.01 eV to 10 MeV. Note that both scales on the graphs are logarithmic. It is useful to divide the energy range into three parts and pick out the features of particular interest. At very low energies, below 0.1 eV, the $(1/v)$ law is clearly seen in the ^{235}U total and fission cross-sections, and the cross-sections are large, due to an excited state of ^{236}U lying just below $E=0$. The fission fraction σ_f/σ_{tot} is $\approx 84\%$; the remaining 16% of σ_{tot} corresponds mostly to radiative capture (the formation of ^{236}U with γ-ray emission). In contrast, the cross-section for ^{238}U is very much smaller and nearly constant in this region, and is due almost entirely to elastic scattering.

The second region is that between 1 eV and 1 keV, where resonances are prominent in both isotopes. These resonances are very narrow and radiative capture gives a significant fraction of the total widths. This is particularly true of the resonances in ^{238}U, which are below the fission threshold in this region; for example, γ-decays account for 95% of the width of the resonance at 6.68 eV.

In the third region, between 1 keV and 3 MeV, the resonances are not resolved in the measured cross-sections. Compound nuclear states at these energies are more dense and wider. Thus the probability of radiative capture is, on average, smaller than at lower energies. The fission cross-section for ^{238}U appears above 1.4 MeV and the ^{235}U fission fraction σ_f/σ_{tot} remains significant. However, in both isotopes at these higher energies the result of a neutron interaction is predominantly scattering, either elastic scattering, or at higher energies, inelastic scattering with neutron energy lost in exciting the nucleus. (The threshold energies for inelastic scattering in ^{235}U and ^{238}U are 14 keV and 44 keV respectively, which are the energies of the first excited states in these nuclei.) Figure 9.1 shows that the ^{235}U and ^{238}U total cross-sections become similar, around 7 barns, at 3 MeV.

9.1 Total cross-section σ_{tot} and fission cross-section σ_f as a function of energy for neutrons incident on (*a*) ^{235}U, (*b*) ^{238}U. In the region of the dashed lines the resonances are too close together for the experimental data to be displayed on the scale of the figures. Note that both the horizontal and vertical scales are logarithmic. (Data from Garber, D. I. & Kinsey, R. R. (1976), *Neutron Cross Sections*, vol. II, Upton, New York: Brookhaven National Laboratory.)

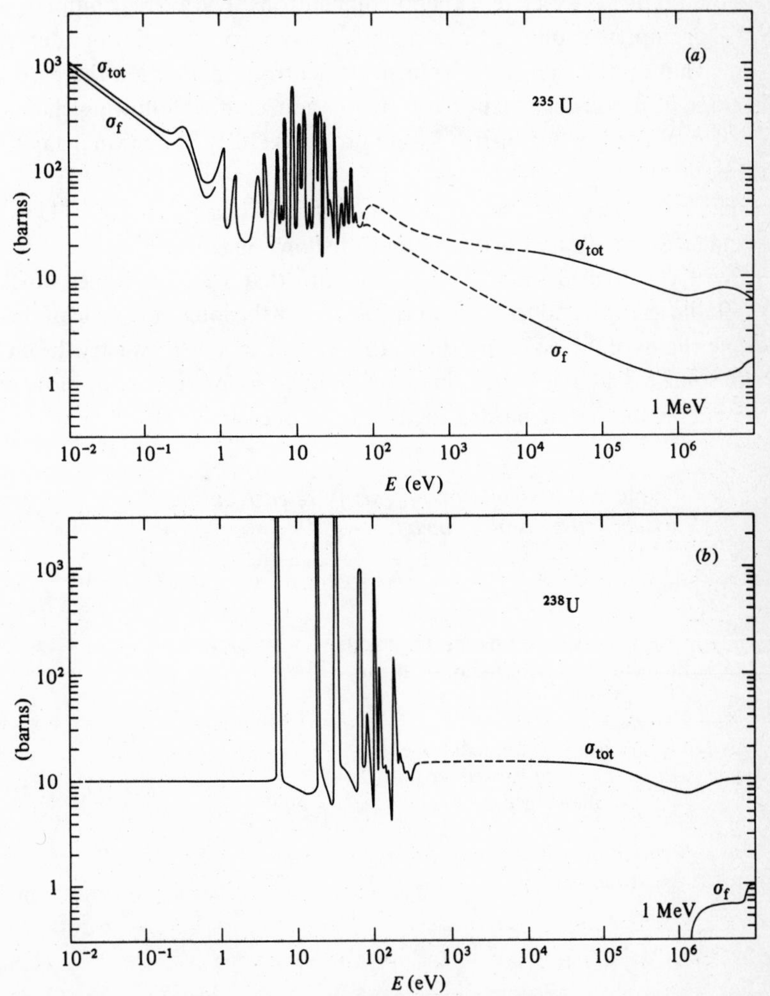

9.3 The fission process

The measured widths of the low-energy resonances in the ^{235}U cross-section are ~ 0.1 eV. The compound nuclei formed at these resonances decay predominantly by fission. Thus we can infer that fission takes place in a time of the order of

$$\tau_f = \frac{\hbar}{\Gamma_f} \approx 10^{-14} \text{ s}$$

after neutron absorption (at least at low energies). On the time scales relevant to this chapter we can regard this as instantaneous.

As with spontaneous fission, there are generally two highly-excited fission fragments which quickly boil off neutrons. The average number ν of these prompt neutrons, per fission, in ^{235}U is $\nu \approx 2.5$. The value of ν depends somewhat on the energy of the incident neutron. In addition there are an average $\delta\nu \approx 0.02$ delayed neutrons per fission, emitted following chains of β^--decays of the neutron-rich fission products (§ 6.5). The mean delay time is about 13 s.

The total energy release on the induced fission of a nucleus of ^{235}U is, on average, 205 MeV and is distributed as shown in Table 9.1.

We have divided the energy release into that which becomes quickly available as heat, and that which is delayed by the long time scale of the β-decay chains of the fission products. In the nuclear power industry the latter is to some extent a nuisance. Some of it is delayed for decades or more and presents a biological hazard in discarded nuclear waste. That which is

Table 9.1. *Distribution of energy release on the induced fission of a nucleus of* ^{225}U

		MeV
Kinetic energy of fission fragments		167
Kinetic energy of fission neutrons		5
Energy of prompt γ-rays		6
		—
Sub-total of 'immediate' energy		178
Electrons from subsequent β-decays	8	
γ-rays following β-decays	7	
	—	
Sub-total of 'delayed' energy		15
Neutrino energy		12
		—
		205
		—

emitted during the lifetime of a fuel-rod is converted into useful heat, but also presents a problem in reactor safety since there is no way of controlling it or turning it off, for example in the case of a breakdown in the heat transport system. In the steady-state operation of a nuclear reactor we shall see that $(\nu-1)$ of the fission neutrons must be absorbed in a non-fission process somewhere in the reactor. Their radiative capture will yield a further 3–12 MeV of useful energy in emitted γ-rays, which is not included in the table. As for the neutrinos, their subsequent interaction cross-sections are so small that almost all of their 12 MeV escapes unimpeded into outer space.

9.4 The chain reaction

Since neutron-induced fission leads to neutron multiplication, in an assembly of uranium atoms there is clearly the possibility of a chain reaction, one fission leading to another or perhaps several more.

Let us first consider some of the length and time scales relevant to a possible chain reaction in uranium metal, which we consider to be a mixture of ^{235}U and ^{238}U atoms in the ratio $c:(1-c)$. The nuclear density ρ_{nuc} of uranium metal is 4.8×10^{28} nuclei m^{-3}.

The average neutron total cross-section for a mixture of the two isotopes is

$$\bar{\sigma}_{\text{tot}} = c\sigma_{\text{tot}}^{235} + (1-c)\sigma_{\text{tot}}^{238}$$

and the mean free path of a neutron in the mixture is

$$l = 1/\rho_{\text{nuc}}\bar{\sigma}_{\text{tot}}$$

(cf. Appendix A). l is the mean distance a neutron travels between interactions. For example, the average energy of a prompt neutron from fission is 2 MeV, and at this energy we see from Fig. 9.1 that $\sigma_{\text{tot}}^{235} \approx \sigma_{\text{tot}}^{238} \approx$ 7 barns. Thus $l \approx 3$ cm. A 2 MeV neutron travels this distance in 1.5×10^{-9} s.

The conceptually most simple case is that of an 'atomic bomb' in which the explosive is uranium highly enriched in ^{235}U. For simplicity we take $c=1$, corresponding to pure ^{235}U. Figure 9.1 shows that a 2 MeV neutron has a 18% chance of inducing fission in an interaction with a ^{235}U nucleus. Otherwise, neglecting the small capture probability at this energy, it will scatter from the nucleus, losing some energy in the process, so that the cross-section for a further reaction may be somewhat increased. If the neutron is not lost from the surface of the metal, the probable number of collisions before it induces fission is about six (Problem 9.4). Assuming the neutron's path is a 'random walk', it will move a net distance of about

$\sqrt{6} \times 3$ cm ≈ 7 cm from its starting point, in a mean time $t_p \approx 10^{-8}$ s, before inducing a further fission and being replaced by, on average, 2.5 new 2 MeV neutrons.

Not all neutrons will induce fission. Some for example will escape from the surface and some will undergo radiative capture. If the probability that a newly-created neutron induces fission is q then each neutron will on average lead to the creation of $(\nu q - 1)$ additional neutrons in the time t_p. (We can neglect delayed neutrons in the present discussion.) If there are $n(t)$ neutrons present at time t, then at time $t + \delta t$ there will be

$$n(t + \delta t) = n(t) + (\nu q - 1)n(t)(\delta t / t_p).$$

In the limit of small δt this gives

$$\frac{\mathrm{d}n}{\mathrm{d}t} = \frac{(\nu q - 1)}{t_p} n(t),$$

which has the solution

$$n(t) = n(0)e^{(\nu q - 1)t/t_p}. \tag{9.1}$$

The number increases or decreases exponentially, depending on whether $\nu q > 1$ or $\nu q < 1$. For ^{235}U the number increases exponentially if $q > (1/\nu) \approx 0.4$.

Clearly, for a small piece of ^{235}U with linear dimensions much less than 7 cm there will be a large chance of escape, q will be small, and the chain reaction will damp out exponentially. However, a sufficiently large mass of uranium brought together at $t = 0$ will have $q > 0.4$. There will be neutrons present at $t = 0$ arising from spontaneous fission and, since $t_p = 10^{-8}$ s, a devastating amount of energy will be released even in a microsecond, before the material has time to disperse. For a bare sphere of ^{235}U the *critical radius* at which $\nu q = 1$ is about 8.7 cm and the *critical mass* is 52 kg.

9.5 Nuclear reactors

We now consider the fate of a 2 MeV neutron in a mass of natural uranium ($c = 0.0072$). It is possible for a 2 MeV neutron to induce fission in either of the two isotopes, but since σ_{tot}^{235} and σ_{tot}^{238} are nearly equal at this energy, the neutron is most likely to interact with ^{238}U which makes up more than 99% of natural uranium. In an interaction with ^{238}U, the probability of fission is only about 5% of that of scattering, which is the predominant interaction in this energy range. Because the uranium nucleus is so massive, the neutron would lose only a small proportion of its energy if it were to scatter elastically (Problem 9.5(*a*)). However, a 2 MeV neutron is likely to scatter inelastically, leaving the ^{238}U nucleus in an excited state, and after one or two such scatterings the neutron's energy will lie below the threshold for inducing fission in ^{238}U.

Once its energy lies below the ^{238}U fission threshold, the neutron has to find a ^{235}U nucleus if it is to induce fission. Its chances of doing this are small unless and until it has 'cooled down' to the very low energies, below 0.1 eV, where the ^{235}U cross-section is much larger than that of ^{238}U (Fig. 9.1). In fact, before the neutron has lost so much energy it is likely to have been captured into one of the ^{238}U resonances, and to have formed the nucleus ^{239}U with the emission of γ-rays. In natural uranium the proportion of fission neutrons which induce further fission is far too small to sustain a chain reaction, even in an infinite mass.

Basically, two routes have been followed to circumvent these difficulties in producing a controlled chain reaction in uranium. The most highly developed technology is that of *thermal reactors*, some of which are fuelled by natural uranium. In a thermal reactor, uranium metal, or more usually the ceramic uranium dioxide, is contained in an array of fuel elements which are in the form of thin rods. Fission neutrons, while still energetic, can escape from the rods into a surrounding large volume filled with material of low mass number and low neutron-absorption cross-section, called the *moderator*. In the moderator the neutrons lose their energy by elastic collisions (Problem 9.5(*b*)) and the volume of the moderator is made so large that a high proportion of the neutrons reach thermal energies corresponding to the ambient temperature of the reactor (0.1 eV = 1160 K). These thermal neutrons, if captured in the fuel rods, are predominantly captured by ^{235}U nuclei, the large cross-section of ^{235}U at thermal energies compensating for its low number-density. Since the neutrons slow down to thermal energies principally in the moderator rather than in the fuel rods, capture into the ^{238}U resonances is largely avoided. The captures into ^{235}U lead to fission with a probability of $\sigma_f^{235}/\sigma_{tot}^{235} \approx 84\%$ at thermal energies, and a chain reaction may be sustained in the reactor in this way. The moderator used in reactors fuelled by natural uranium is ^{12}C in the form of graphite, or 'heavy water', D_2O.

The design criteria of thermal reactors are less stringent if the fuel is artificially enriched with ^{235}U; the reactor can be smaller and it becomes possible to use ordinary water rather than D_2O as a moderator, despite the relatively high neutron-absorption cross-section of hydrogen through the reaction $n + p \rightarrow {}^2H + \gamma + 2.33$ MeV. Typical enrichment in commercial reactors is 2%–3%.

The alternative to the thermal reactor is the *fast reactor*. In a fast reactor there is no large volume of moderator and no large density of thermal neutrons is established. Fission is induced by fast neutrons – hence the name. A fast reactor works because the fission probabilities within the fuel are increased over those of natural uranium by increasing the proportion of fissile nuclei to $\approx 20\%$. The fissile fuel used is ^{239}Pu rather than ^{235}U, for reasons we shall discuss in §9.7.

9.6 Reactor control and delayed neutrons

In a nuclear explosion the delayed neutrons are of no consequence: they appear after the event. In a power reactor they must be considered, since fuel rods can remain in the reactor for three or four years. Thus in a reactor each fission leads to $[(\nu+\delta\nu)q-1]$ additional neutrons, where $\delta\nu$ is the number of delayed neutrons per fission.

In the steady operation of a reactor, with a constant rate of energy production, the neutron density must remain constant so that the reaction rate remains constant. Thus q must be such that the critical condition

$$(\nu+\delta\nu)q-1=0$$

is satisfied.

Reactors are controlled by manipulating q mechanically, using adjustable control rods inserted in the reactor. The control rods contain materials such as boron or cadmium, which have a large neutron-absorption cross-section in the thermal energy range (Fig. 8.3). Inserting or withdrawing the control rods decreases or increases q. It is important in the design of reactors that the critical condition cannot be met by the prompt neutrons alone, so that

$$\nu q-1<0$$

always. Although the lifetime of a prompt neutron in a thermal reactor may be as long as 10^{-3} s, rather than 10^{-8} s which we estimated in pure ^{235}U, this gives an uncomfortably short time scale in which to change q mechanically and so avoid an accidental catastrophic exponential rise in neutron density, as given by equation (9.1). However, since the reactor can only become critical for

$$(\nu+\delta\nu)q-1=0$$

the time scale becomes that of the delayed neutrons and the response time adequate for mechanical control.

A reactor is brought into operation by increasing q and allowing the neutron density to increase until the required power production and operating temperature is reached. The heat produced, to be used in the more traditional technology of raising steam and driving turbines, is carried away by a coolant circulating through tubes which permeate the core of the reactor, to a heat exchanger outside the reactor (Fig. 9.2). Thus the coolant is, necessarily, also a moderator, and its nuclear properties as well as its thermal properties have to be considered. Gas-cooled thermal reactors have commonly used carbon dioxide under pressure. Ordinary water can be used as coolant in reactors using enriched uranium, such as the American and French PWRs (pressurised-water reactors, in which the water is kept under pressure to prevent it boiling). In the case of a fast reactor, the absence of moderator necessitates a highly compact core which demands a coolant of high thermal conductivity and high thermal capacity; liquid sodium appears to be most suitable and has been used in prototype reactors.

For thermal stability, it is very important that q, the proportion of neutrons inducing fission, satisfies

$$\frac{\mathrm{d}q}{\mathrm{d}T} < 0,$$

so that an increase in temperature T leads to a fall in q, and hence a fall in the reaction rate and *vice-versa*. There are many factors affecting $\mathrm{d}q/\mathrm{d}T$, arising from the thermal expansion of the various components of the reactor, changes in the velocity distribution of the thermal neutrons with temperature, and the effect of Doppler broadening of resonances. In thermal reactors, Doppler broadening leads to an increase in the neutron absorption in ^{238}U resonances in the fuel rods and gives a significant negative contribution to $\mathrm{d}q/\mathrm{d}T$. Since the resonant cross-sections are large, neutrons which impinge upon fuel rods and whose energies lie near to resonances are absorbed close to the surface of the rod. The broadening of the resonance increases the energy band absorbed and hence increases the neutron absorption rate.

In a fast reactor the effects of Doppler broadening are more complicated since the fission rate in ^{239}Pu resonances is also increased by broadening. It is important for the safety of fast reactors that the net effect on $\mathrm{d}q/\mathrm{d}T$ should be negative.

9.7 Production and use of plutonium

So far we have considered only ^{235}U as a nuclear fuel and regarded ^{238}U with its high radiative-capture cross-section as something of a drawback. However, the nucleus ^{239}U formed in radiative capture is odd

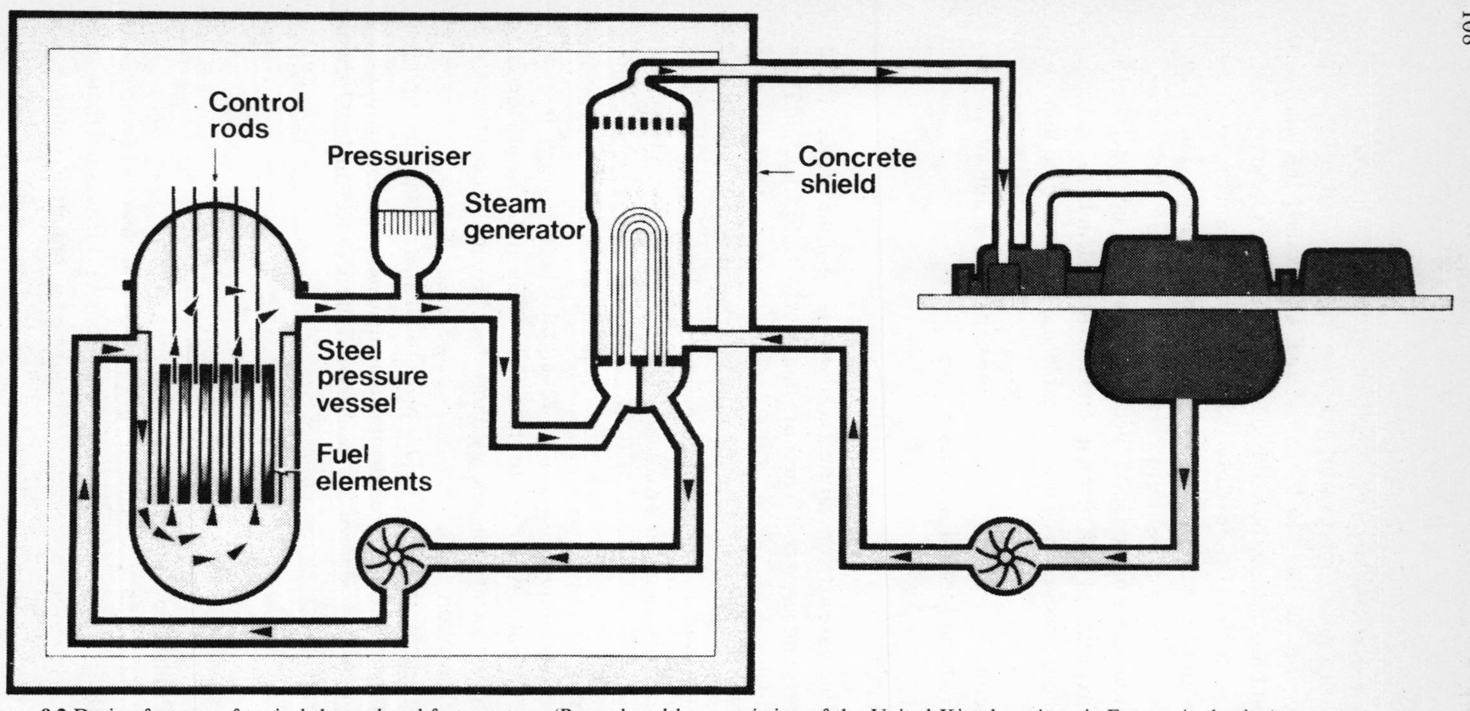

9.2 Design features of typical thermal and fast reactors. (Reproduced by permission of the United Kingdom Atomic Energy Authority.)
(*a*) Thermal reactor – water moderated. **Fuel** Uranium dioxide clad in an alloy of zirconium (Zircaloy). **Moderator** Light water (ordinary water, H_2O). **Core layout** Fuel pins, arranged in clusters, are placed inside a pressure vessel containing the light water moderator, which also is the coolant. **Heat extraction** The light water in the pressure vessel at high pressure is heated by the core. It is pumped to a steam generator where it boils water in a separate circuit; the steam drives a turbine coupled to an electric generator. **Indicative data for a reactor of 1200 MW(e) size** *Uranium enrichment* (% ^{235}U) 3.2%, *Coolant outlet temperature* 324°C, *Coolant pressure* 2250 psia, *Steam cycle efficiency* 32%, *Core dimensions* 3.0 m dia. × 3.7 m high

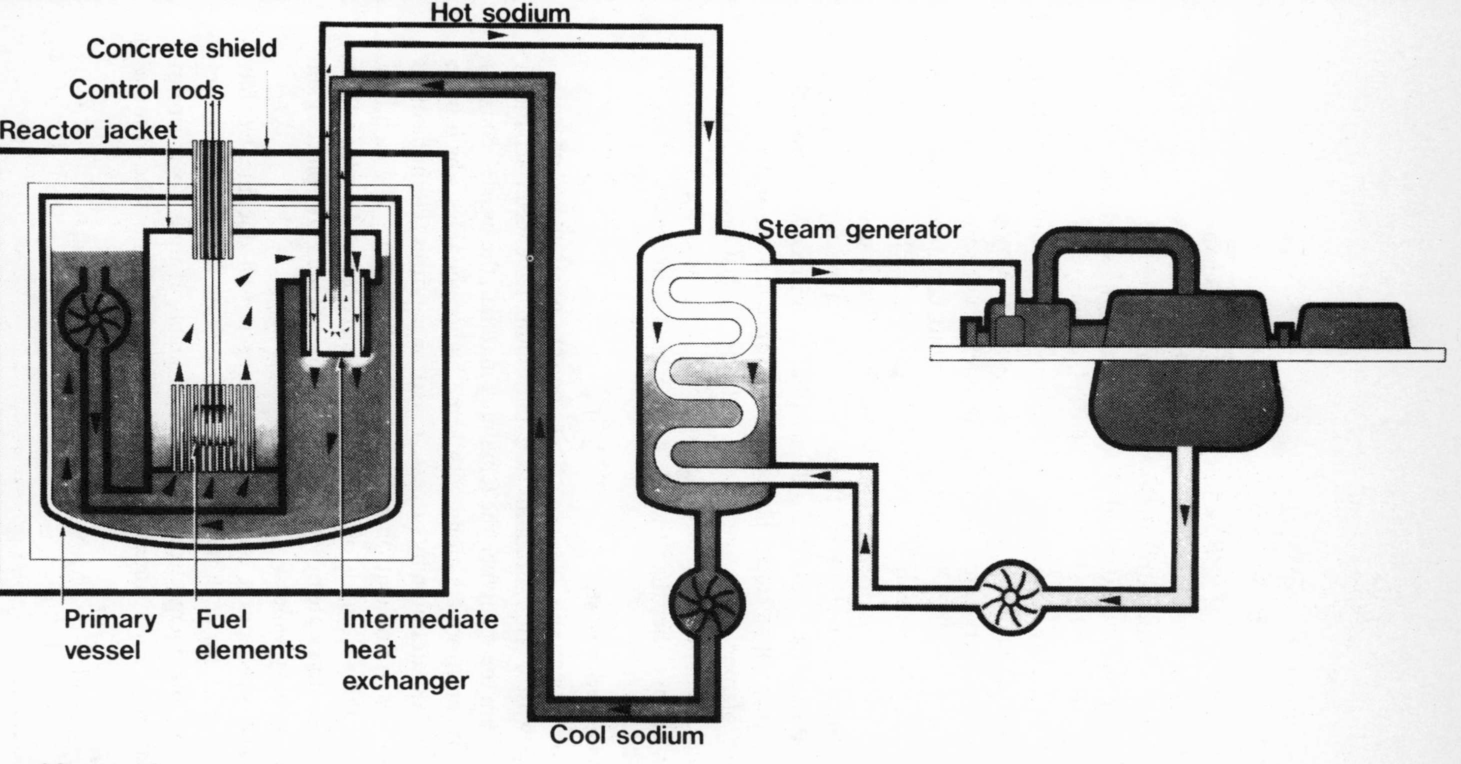

9.2 *continued*

(*b*) Fast reactor – sodium cooled. **Fuel** A mixture of plutonium and uranium dioxides in stainless steel cans. **Moderator** None. **Core layout** Assemblies of fuel elements are placed inside a tank containing the liquid sodium coolant. The core is surrounded by a 'blanket' of depleted uranium (dioxide) in stainless steel cans. **Heat extraction** The sodium is heated by the core and pumped through an intermediate heat exchanger where it heats sodium in a separate secondary circuit. The sodium in the secondary circuit transfers its heat to water in a steam generator; the steam drives a turbine coupled to an electric generator. **Indicative data for a reactor of 1300 MW(e) size** *Fuel enrichment* (% Pu) 20%, *Coolant outlet temperature* 620°C, *Coolant pressure* 5 psig, *Steam cycle efficiency* 44%, *Core dimensions* 2.3 m dia. × 1.1 m high, *Core and blanket dimensions* 3.1 m dia. × 2.1 m high and β-decays to the fissile nucleus ^{239}Pu:

$$^{239}_{92}\text{U} \xrightarrow[(34\ \text{min})]{} {}^{239}_{93}\text{Np} + e^- + \bar{\nu}$$

$$\downarrow (3.36\ \text{days})$$

$$^{239}_{94}\text{Pu} + e^- + \bar{\nu}.$$

The nuclear properties of ^{239}Pu are very similar to ^{235}U and, in particular, it is suitable as a fuel in a nuclear reactor. In a thermal reactor, some of the ^{239}Pu produced will be burnt up in the lifetime of the fuel rods, and the remainder later may be extracted chemically from the spent fuel.

Because of the relatively short mean life of plutonium isotopes (^{239}Pu has a mean life for α-decay of 3.5×10^4 years) virtually all plutonium on Earth is man-made. Large quantities have been produced as a by-product of the nuclear power industry (and wilfully for the nuclear weapons programme).

The value of ν for ^{239}Pu is 2.96 for fast neutrons, compared with 2.5 for ^{235}U, so that it is a very suitable fuel for fast reactors. Such reactors can be designed to breed more fissile ^{239}Pu from ^{238}U than is consumed, using 'spare' neutrons. In a fast reactor the central core is, typically, loaded with 20% of ^{239}Pu and 80% of ^{238}U ('depleted' uranium recovered from the operation of thermal reactors). The core is enveloped in a 'blanket' of ^{238}U, and in this blanket more plutonium is made. A fast-breeder reactor programme can, in principle, be designed to utilise all the energy content of natural uranium, rather than the 1% or so exploited in thermal reactors.

9.8 Radioactive waste

The operation of a nuclear power programme generates radioactive waste. After uranium and plutonium have been separated chemically from the spent fuel, the remaining material, the 'waste', consists mainly of fission products along with some higher actinides which have been built up from uranium by successive neutron captures. The immediate products of fission are neutron rich, and hence β-emitters. The daughter nucleus from the β-decay is often formed in an excited state, which then decays to its ground state by γ-emission. β-decay will then take place again until the β-stability valley is reached.

A complete description of the decay chains is well documented but complex. Overall, for each fission it is found that on average the rate of release of ionising energy from the decay products at time t is given, to within a factor of 2, for times between 1 s and 100 years by the formula

$$\frac{dE}{dt} = 2.66\left(\frac{1\,s}{t}\right)^{1.2}\ \text{MeV s}^{-1}. \tag{9.2}$$

Over this period the energy release is divided roughly equally between electrons and γ-rays. Energy lost to neutrinos is not included. Problem 9.8 indicates how such a simple empirical formula can be used to estimate properties such as heat output and radioactivity of the waste.

Problems

9.1 The combustion of methane

$$CH_4 + 2O_2 \rightarrow CO_2 + 2H_2O,$$

releases an energy of about 9 eV/(methane molecule). Estimate the relative energy release per unit mass for nuclear (fission) as against chemical fuels.

9.2 Show that a nuclear power plant producing 1000 MW of heat consumes about 1 kg of ^{235}U (or other fissionable fuel) per day.

9.3 Show that the semi-empirical mass formula predicts that for a heavy nucleus the neutron separation energy (or 'binding energy of the last neutron') is approximately $2 \times (11.2/A^{\frac{1}{2}})$ MeV greater for an even Z even N nucleus such as ^{236}U than it is for a nearby even Z odd N nucleus such as ^{239}U.

9.4 Suppose that a neutron induces fission in a nucleus with probability p, and that otherwise the collision is elastic. Show that the mean number of collisions it undergoes is $1/p$.

9.5(*a*) A neutron with kinetic energy T_0 (non-relativistic) collides elastically with a stationary nucleus of mass M. In the centre-of-mass system the scattering is isotropic. Show that on average the neutron energy after the collision is

$$T_1 = \frac{M^2 + m_n^2}{(M + m_n)^2} T_0.$$

(*b*) Consider the nuclei of a graphite moderator to be pure ^{12}C, with a number density of 0.9×10^{29} nuclei/m^3. For neutron energies less than 2 MeV the scattering is elastic, with a cross-section approximately constant $\sim$4.5 b.

Estimate (*a*) the number of collisions required to reduce the energy of a 2 MeV fission neutron to a thermal energy of 0.1 eV, and (*b*) the time it takes.

9.6 If the neutron density $\rho(\mathbf{r}, t)$ in a material is slowly varying over distances long compared with the neutron mean free path l, $\rho(\mathbf{r}, t)$ approximately satisfies the 'diffusion equation with multiplication',

$$\frac{\partial \rho}{\partial t} = \frac{(\nu - 1)}{t_p} \rho + D \nabla^2 \rho.$$

The coefficient of diffusion is given in simple transport theory by $D=lv/3$, where v is the neutron velocity (assumed constant). At a free surface, the effective boundary condition, again obtained from transport theory, is

$$0.71l\frac{\partial\rho}{\partial n}+\rho=0,$$

where $\partial/\partial n$ denotes differentiation along the outward normal to the surface.

Using the data given in §9.4, estimate the critical radius of a bare sphere of ^{235}U. Look for spherically symmetric solutions of the equation of the form $\rho(r,t)=f(r)e^{\lambda t}$, and replace the boundary condition at the surface $r=R$ by the approximation $\rho(R+0.71l,t)=0$.

9.7 In a simplified model, the number of neutrons $n(t)$ in a reactor at time t is given by

$$\frac{dn}{dt}=\frac{(vq-1)}{t_p}n+\frac{\delta vq}{t_p}\int_{-\infty}^{t}\frac{n(t')e^{-(t-t')/\tau_\beta}}{\tau_\beta}dt',$$

where t_p is the mean life of the prompt neutrons and τ_β is the mean life of those fission fragments which produce delayed neutrons.

(*a*) Show how this equation may be derived, assuming that only one type of fission fragment produces delayed neutrons.

(*b*) Show that solutions are of the form

$$n(t)=n_0e^{\lambda t},$$

and give the equation for λ.

(*c*) Show that if $t_p=10^{-4}$ s, $vq-1=10^{-4}$, and there are no delayed neutrons ($\delta v=0$), then $n(t)$ increases exponentially with a time-scale of 1 s.

(*d*) Show that if $\tau_\beta=10$ s, $(v+\delta v)q-1=10^{-4}$ and $vq-1=-0.0078$ (corresponding to $v=2.5$, $\delta v=0.02$), $n(t)$ increases exponentially with a time-scale of about 13 minutes.

9.8(*a*) Show that the mean thermal power from a fuel rod of a reactor that has been shut down for time t (>1 s), after burning with steady power output P for a time T, is approximately

$$\text{power}=0.07P\left[\left(\frac{1\text{ s}}{t}\right)^{0.2}-\left(\frac{1\text{ s}}{T+t}\right)^{0.2}\right]$$

(*b*) Before its catastrophic shut-down, the No. 4 Chernobyl reactor had been producing about 3 GW of heat. Taking the mean age of its fuel rods to be $T=1$ year, estimate the power outputs from the core at one week, one month, and one year after the accident. 97% of the radioactive material remained in the core. (Answer: 8.0 MW, 4.4 MW, 0.9 MW.)

10
Nuclear fusion

In this chapter we describe the nuclear reactions that power the Sun and thus make possible life on Earth. In contrast to the power from fission discussed in the preceding chapter, the radiance of the Sun comes from the fusion of the lightest element, hydrogen, into helium. We then examine the possibility of controlled nuclear fusion for power production on Earth.

10.1 The Sun

In stars, the gravitational, the weak, the electromagnetic, and the strong interactions all play an active and essential role. Our Sun and its planets are thought to have condensed out of a diffuse mass of material, mostly hydrogen and helium atoms, some 4.6×10^9 years ago. Table 10.1 gives the estimated proportions of the ten most abundant nuclei in that mass of material.

The major attributes of the Sun, determined from a wide variety of observations, are as follows:

Mass $\quad M_\odot = 1.99 \times 10^{30}$ kg

Radius $\quad R_\odot = 6.96 \times 10^8$ m

Luminosity $\quad L_\odot = 3.86 \times 10^{26}$ W.

(The luminosity of a star is the total rate of emission of electromagnetic energy.)

Because of the long range and universally attractive nature of gravity, a homogeneous mass of gas at sufficiently low temperature is unstable to contraction into objects like stars. During contraction of a mass of gas, gravitational potential energy is converted into kinetic energy and radiation energy, and the temperature of the gas rises. The rate of collapse is determined by the extent to which the build-up of pressure in the hot, dense interior can balance the incessant pressure of gravitational contraction. In a star like the Sun, as the temperature and density increased, its rate of contraction was essentially stopped when the interior became hot enough to ignite the hydrogen-burning reaction that we shall discuss in detail presently. At this stage in the Sun's evolution, the generated nuclear power keeps the interior hot enough to sustain the pressure that balances gravity, and a quasi-static condition is established, a condition that exists today. This condition is not one of thermodynamic equilibrium, since the interior is hotter than the outside and the nuclear energy liberated at the centre is transferred, radiatively and by conduction and convection, to the surface, where it is radiated out into space, to our benefit, and gives the Sun its luminosity.

The principal reactions that power the Sun begin with the conversion of hydrogen into deuterium:

$$\mathrm{p}+\mathrm{p} \to {}^{2}_{1}\mathrm{H}+\mathrm{e}^{+}+\nu+0.42\ \mathrm{MeV}. \tag{10.1}$$

This reaction involves the weak interaction (a proton changes to a neutron), and so occurs very rarely. It is the weak interaction that sets the long time scale of the quasi-static state of the Sun.

The positron produced in the reaction quickly annihilates with an electron to release a further 1.02 MeV of energy. The deuterium is converted to $^{3}\mathrm{He}$:

$$\mathrm{p}+{}^{2}_{1}\mathrm{H} \to {}^{3}_{2}\mathrm{He}+\gamma+5.49\ \mathrm{MeV}, \tag{10.2}$$

which in turn fuses to $^{4}\mathrm{He}$:

$${}^{3}_{2}\mathrm{He}+{}^{3}_{2}\mathrm{He} \to {}^{4}_{2}\mathrm{He}+\mathrm{p}+\mathrm{p}+12.86\ \mathrm{MeV}. \tag{10.3}$$

Table 10.1. *The proportion by number, relative to carbon, of the ten most abundant atoms in the Solar System at its birth*

H	He	C	N	O	Ne	Mg	Si	S	Fe
2400	162	1.0	0.21	1.66	0.23	0.10	0.09	0.05	0.08

Data from Cameron, A. G. W. (1982), in *Essays in Nuclear Astrophysics*, ed. C. A. Barnes, D. D. Clayton & D. N. Schramm. Cambridge University Press, p. 23.

Thus the net result of these reactions, which are called the 'PPI chain', is the conversion of hydrogen to helium with an energy release of 26.73 MeV per helium nucleus formed. The neutrinos emitted in the p–p reactions take an average 0.26 MeV of energy each. This energy is lost into outer space, but is not included in the observed luminosity. Thus each hydrogen atom consumed in this process leads to the emission of 6.55 MeV of electromagnetic energy from the Sun.

The observed solar luminosity implies that $L_\odot/(6.55\ \mathrm{MeV}) = 3.7 \times 10^{38}$ hydrogen atoms are converted into helium per second. This rate of conversion, over the lifetime of the Sun, gives a total of 5.4×10^{55} conversions. The Sun's mass and composition show that it started with about 8.9×10^{56} hydrogen atoms. We can conclude that less than 10% of the hydrogen of the Sun has so far been consumed, and appreciate the long time scale of this stage of stellar evolution.

Figure 10.1 shows the density, temperature and thermonuclear-power density from a model calculation of the Sun as it is now. It is interesting to note that 50% of the mass is within a distance of $R_\odot/4$ from the centre, and 95% of the luminosity is produced in the central region within a distance of $R_\odot/5$, where the temperature is such that $k_B T \gtrsim 1$ keV. (k_B is Boltzmann's constant and $k_B T = 1$ keV when $T = 1.16 \times 10^7$ K.)

A simple order-of-magnitude calculation shows that the gravitational energy released in contraction, before the quasi-static period began, is sufficient to produce such temperatures:

$$\text{gravitational energy} \sim \frac{GM_\odot^2}{R_\odot} = 3.8 \times 10^{41}\ \mathrm{J},$$

where the gravitational constant $G = 6.67 \times 10^{-11}\ \mathrm{m^3\,kg^{-1}\,s^{-2}}$. This energy would give ~ 1 keV of kinetic energy on average to every particle, nuclei and electrons. At these temperatures the hydrogen and helium atoms will be completely ionised; material in this condition is known as *plasma*.

10.2 Cross-sections for hydrogen burning

We turn now to a more-detailed examination of the nuclear physics involved in hydrogen burning. Reactions involving charged particles were discussed in § 8.3, where the role of the Coulomb barrier was emphasised. For energies in the range of keV we shall use the low-energy expression for charged-particle reactions given by equation (8.7):

$$\sigma(E) = \frac{1}{E} S(0) e^{-\sqrt{(E_G/E)}} \tag{10.4}$$

where

$$E_G = 2mc^2\left(\frac{\pi Z_1 Z_2 e^2}{\hbar c(4\pi\varepsilon_0)}\right)^2.$$

A more accurate expression may be necessary if there is a resonance in the keV energy range.

Direct measurements of the cross-sections of reactions (10.2) and (10.3) at energies of 1 keV have not been made since they are so small, but values of

10.1 (*a*) Mass densities and (*b*) the thermonuclear power density ε and the temperature *T*, in the modern Sun as a function of distance *r* from the centre. (Taken from model calculations by Bahcall, J. N. *et al.* (1982), *Rev. Mod. Phys.* **54**, 767.)

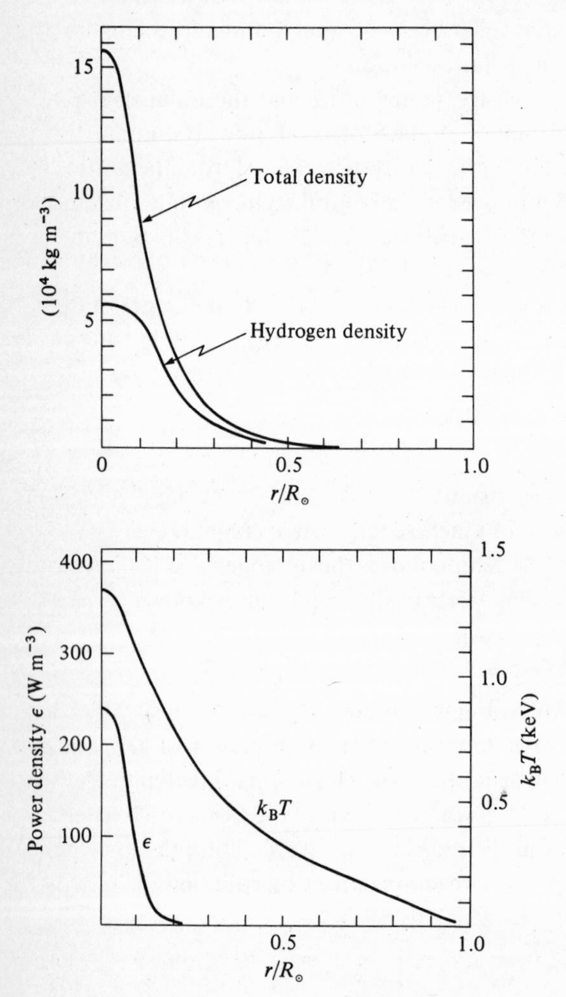

$S(0)$ are known by extrapolation from measurements at higher energies. The p–d reaction (10.2) has been measured down to 15 keV and $S_{pd}(0) \approx 2.5 \times 10^{-7}$ MeV b. The helium–helium reaction (10.3) has been measured down to 33 keV, giving $S_{hh}(0) \approx 4.7$ MeV b. The p–d cross-section is small because it involves a γ-transition. The p–p reaction (10.1), the first stage of hydrogen burning, has a cross-section which is lower by many orders of magnitude because it involves the weak interaction. Although the reaction is crucial, the cross-section is so small that it has not been directly measured in a laboratory at any energy. Fortunately we have such a precise theory of the weak interaction that the cross-section, and $S_{pp}(0)$, can be calculated with some confidence; it is found that

$$S_{pp}(0) \approx 3.88 \times 10^{-25} \text{ MeV b}. \qquad (10.5)$$

This order of magnitude is not too difficult to understand. As is explained in Chapter 3, the proton–proton nuclear potential has been determined from scattering experiments. From the potential, the low-energy proton–proton nuclear-scattering cross-section can be calculated. Although there is no resonance, the nuclear attraction makes the cross-section quite large, 36 barns, at energies $\gtrsim 1$ MeV which are low but are above the Coulomb barrier. Reaction (10.1) involves the protons coming together within the range of the nuclear force (Fig. 3.2) and, while they are together, a β-decay taking place. We can estimate the probability of this β-decay by

$$\text{probability} = \frac{\text{a typical nuclear time}}{\text{a } \beta\text{-decay time}}.$$

Consider the cross-section at 1 MeV. Since the energy released in the reaction is comparable with the energy released in the β-decay of a free neutron, it is reasonable to take the β-decay time to be the neutron lifetime, 898 s, and the nuclear time is $\sim 10^{-23}$ s (§ 5.2). The cross-section for the proton–proton to deuteron reaction at 1 MeV should thus be of the order

$$\sigma \sim (36 \text{ b}) \times \left(\frac{10^{-23} \text{ s}}{898 \text{ s}} \right) = 4 \times 10^{-25} \text{ b}.$$

Since the energy of 1 MeV is above the Coulomb barrier, we can infer that $S_{pp} \sim 4 \times 10^{-25}$ MeV b. This excellent agreement with equation (10.5) is fortuitous, but the argument does make intelligible the order of magnitude of this key quantity.

10.3 Nuclear reaction rates in a plasma

During the early cold stages of stellar contraction, nuclei do not have kinetic energies high enough, compared with the Coulomb barriers between them, for the barrier penetration probability to be significant. To obtain the reaction rate for a process in the interior of a star and see how it depends on temperature, we must average suitably over the energies of the particles involved. The calculation is an important one so we shall set it out here.

Consider the nuclei in a volume of plasma small enough for the temperature and number densities to be considered uniform. We shall assume that the velocities of the nuclei are given by the Maxwell–Boltzmann distribution, so that the probability of two nuclei having a relative velocity v in the range, v, $v+\mathrm{d}v$ is given by

$$P(v)\,\mathrm{d}v=\left(\frac{2}{\pi}\right)^{\frac{1}{2}}\left(\frac{m}{k_{\mathrm{B}}T}\right)^{\frac{3}{2}}e^{-mv^2/2k_{\mathrm{B}}T}v^2\,\mathrm{d}v,$$

where m is the reduced mass of the pair. (The centre-of-mass motion has been factored out.)

If the nuclei are labelled by a, b, with number densities ρ_a, ρ_b, the number of reactions per unit time per unit volume is

$$\text{reaction rate per unit volume}=K\rho_a\rho_b(\overline{v\sigma_{ab}}), \tag{10.6}$$

where σ_{ab} is the cross-section for the reaction, and the bar denotes the average over the velocity distribution. Equation (10.6) follows from the discussion of reaction rates in Appendix A. The factor $K\rho_a\rho_b$ is the number of interacting pairs in a unit volume of the system; $K=1$ if the nuclei are different and $K=\frac{1}{2}$ if the nuclei are the same.

We have

$$\overline{v\sigma_{ab}}=\int_0^\infty v\sigma_{ab}P(v)\,\mathrm{d}v.$$

Changing variables to $E=mv^2/2$, and using the low-energy formula (10.4) for the cross-section, this becomes

$$\overline{v\sigma_{ab}}=\left(\frac{8}{\pi m}\right)^{\frac{1}{2}}\left(\frac{1}{k_{\mathrm{B}}T}\right)^{\frac{3}{2}}S_{ab}(0)\int_0^\infty e^{-\phi(E)}\,\mathrm{d}E, \tag{10.7}$$

where $\phi(E)=E/k_{\mathrm{B}}T+\sqrt{(E_G/E)}$.

The function $e^{-\phi(E)}$ is sharply peaked. It falls off rapidly at high energies because of the Boltzmann factor, and at low energies because of the barrier-penetration factor. The peak lies at $E=E_0$ where $\phi(E)$ is a minimum, i.e. where $\mathrm{d}\phi/\mathrm{d}E=0$, which gives

$$E_0=(E_G)^{\frac{1}{3}}(k_{\mathrm{B}}T/2)^{\frac{2}{3}}.$$

Figure 10.2 is a graph of $e^{-\phi(E)}$ appropriate to the p–p reaction at the centre of the Sun.

As it stands, the integral cannot be performed analytically, but the main contribution comes from the peak. It is possible to replace $\phi(E)$ in the neighbourhood of E_0 by a simpler expression, leading to an analytic result for the integral (which is also a good approximation). The Taylor expansion of $\phi(E)$ about E_0 gives

$$\phi(E) \approx \phi(E_0) + \tfrac{1}{2}(E-E_0)^2\phi''(E_0),$$

where

$$\phi(E_0) = 3(\tfrac{1}{2})^{\frac{2}{3}}(E_G/k_BT)^{\frac{1}{3}}$$
$$\phi''(E_0) = 3(\tfrac{1}{2})^{\frac{1}{3}}E_G^{-\frac{1}{3}}(k_BT)^{-\frac{5}{3}}$$

The linear term does not appear, since $\phi'(E_0)=0$. With this approximation, the integrand is replaced by a Gaussian peak and the integration range can be extended down to $E=-\infty$ with negligible error. We have then, remembering the result $\int_{-\infty}^{\infty} e^{-ax^2}\,dx = (\pi/a)^{\frac{1}{2}}$, the fairly simple expression

$$\overline{v\sigma_{ab}} = \tfrac{8}{9}S_{ab}(0)\left(\frac{2}{3mE_G}\right)^{\frac{1}{2}}\tau^2 e^{-\tau}$$

with

$$\tau = 3(\tfrac{1}{2})^{\frac{2}{3}}(E_G/k_BT)^{\frac{1}{3}} = 3\left(\frac{mc^2}{2k_BT}\right)^{\frac{1}{3}}\left(\frac{\pi Z_a Z_b e^2}{\hbar c(4\pi\varepsilon_0)}\right)^{\frac{2}{3}}. \qquad (10.8)$$

For practical calculations, taking the masses of nuclei to be $A \times$ (one atomic

10.2 The function $\exp(-\phi(E))$ appearing in equation (10.7), plotted for the proton–proton reaction at $k_BT = 1.34$ keV.

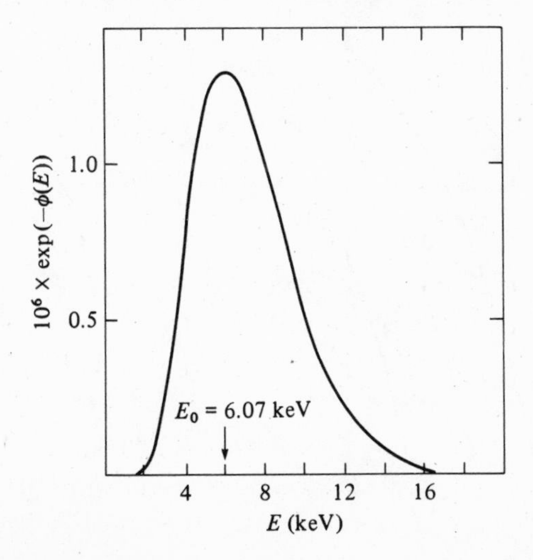

mass unit) gives

$$\overline{v\sigma}_{ab} = \frac{7.21 \times 10^{-22}}{Z_a Z_b} \frac{(A_a + A_b)}{A_a A_b} \left(\frac{S_{ab}(0)}{1 \text{ MeV b}} \right) \tau^2 e^{-\tau} \text{ m}^3 \text{ s}^{-1},$$

and

$$\tau = 18.8 \left(\frac{Z_a^2 Z_b^2 A_a A_b}{A_a + A_b} \right)^{\frac{1}{3}} \left(\frac{1 \text{ keV}}{k_B T} \right)^{\frac{1}{3}}.$$

Note that the temperature dependence of $\overline{v\sigma}$ lies entirely in the factor $\tau^2 e^{-\tau}$.

The temperature dependence is dramatic, as also is the dependence on the nuclear species involved. Both are illustrated in Fig. 10.3, which shows plots of $\tau^2 e^{-\tau}$ against temperature for several reactions of astrophysical interest. Note that the vertical scale extends over a range of 10^{60}!

For a given set of nuclear reactions, we can write down equations giving the rate of change of the number densities ρ of the nuclei participating, in a region of given temperature. For example, considering the PPI reactions and writing $\lambda_{ab} = \overline{v\sigma}_{ab}$, we have, from reactions (10.1) and (10.2) and equation (10.6),

$$\frac{\partial \rho_d}{\partial t} = \tfrac{1}{2} \lambda_{pp} \rho_p^2 - \lambda_{pd} \rho_p \rho_d.$$

10.3 The function $\tau^2 \exp(-\tau)$ appearing in equation (10.8), for some nuclear reactions.

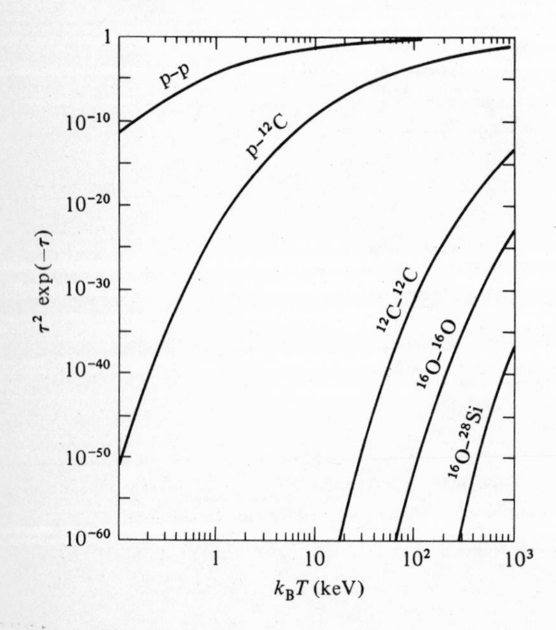

Because of the long time scale of hydrogen burn-up, ρ_p may be regarded as constant in this equation, giving the solution

$$\rho_d(t) = (\tfrac{1}{2}\lambda_{pp}/\lambda_{pd})\rho_p(1 + Ce^{-\lambda_{pd}\rho_p t}),$$

where C is a constant.

Using the numerical values for $S_{pp}(0)$ and $S_{pd}(0)$ given in § 10.2, and a temperature and density appropriate to the centre of the Sun, from Fig. 10.1, we find that the time constant for establishing equilibrium is $(\lambda_{pd}\rho_p)^{-1} = 3.3$ s. Thus our assumption that ρ_p could be treated as a constant was valid. In equilibrium, the ratio $\rho_d/\rho_p = \frac{1}{2}\lambda_{pp}/\lambda_{pd} = 1.5 \times 10^{-18}$. The low density of deuterium accounts for our neglect of d–d reactions (which are considered in § 10.5).

10.4 Other solar reactions, and solar neutrinos

Our account above of hydrogen burning in the Sun is not complete. There are other ways of consuming the ^{3}He formed in reaction (10.2). The presence of ^{4}He in a star leads to the formation of ^{7}Be:

$$^3_2\text{He} + ^4_2\text{He} \rightarrow ^7_4\text{Be} + \gamma + 1.59 \text{ MeV}.$$

^{7}Be is unstable to the capture of a free electron from the plasma to form ^{7}Li:

$$^7_4\text{Be} + \text{e} \rightarrow ^7_3\text{Li} + \nu + 0.86 \text{ MeV},$$

and $^7_3$Li is quickly broken up by a proton into two helium nuclei:

$$^7_3\text{Li} + \text{p} \rightarrow ^4_2\text{He} + ^4_2\text{He} + 17.35 \text{ MeV};$$

this is the 'PPII chain'.

Alternatively, the $^7_4$Be may interact with a proton to form $^8_5$B:

$$^7_4\text{Be} + \text{p} \rightarrow ^8_5\text{B} + \gamma + 0.14 \text{ MeV}.$$

$^8_5$B is unstable to β-decay,

$$^8_5\text{B} \rightarrow ^8_4\text{Be}^* + \text{e}^+ + \nu + 14.02 \text{ MeV},$$

and $^8_4$Be* breaks up into two helium nuclei:

$$^8_4\text{Be}^* \rightarrow ^4_2\text{He} + ^4_2\text{He} + 3.03 \text{ MeV};$$

this is the 'PPIII chain'. The positron annihilates with an electron to release a further 1.02 MeV.

The relative importance of the PPII and PPIII chains, compared with the PPI chain, can be calculated from the appropriate set of rate equations; in the standard model of the Sun, the PPI chain is the main process.

Another interesting set of reactions resulting in the burning of hydrogen to helium is the 'CNO' cycle. The presence of any of the nuclei $^{12}_6$C, $^{13}_6$C, $^{14}_7$N

or $^{15}_{7}N$ catalyses the burning by the set of reactions

$$^{12}C + p \rightarrow {}^{13}N + \gamma, \quad {}^{13}N \rightarrow {}^{13}C + e^{+} + \nu$$
$$^{13}C + p \rightarrow {}^{14}N + \gamma$$
$$^{14}N + p \rightarrow {}^{15}O + \gamma, \quad {}^{15}O \rightarrow {}^{15}N + e^{+} + \nu$$
$$^{15}N + p \rightarrow {}^{12}C + {}^{4}He.$$

The weak interactions in the cycle are not compelled to occur in a fleeting 10^{-23} s, as in the p–p reaction, but can proceed at their leisure in the usual β-decay times. Carbon and nitrogen nuclei are known to be present in the Sun (Table 10.1), but at the temperatures of the Sun the reaction rates are greatly suppressed by the Coulomb barrier (Fig. 10.3), and the CNO cycle probably accounts for only about 3% of stellar hydrogen burning. In hotter stars the CNO cycle may dominate over the PP chains, since the CNO cycle reaction rates increase more rapidly with temperature (Fig. 10.3 again).

All these reactions in the Sun lead to a considerable flux of neutrinos through the Earth. The detection of the high-energy tail of this flux has been attempted through the reaction

$$\nu + {}^{37}_{17}Cl + 0.81 \text{ MeV} \rightarrow e^{-} + {}^{37}_{18}Ar.$$

This reaction is endothermic and the neutrinos from the PPI process do not have the energy to induce it. It is triggered most effectively by the high-energy (up to 14.02 MeV) neutrinos emitted in the ^{8}B β-decay of the PPIII chain.

Neutrino interaction cross-sections, as we have often remarked, are small. The detector in the experiment consists of 615 tons of liquid perchloroethylene, C_2Cl_4, i.e. 2.18×10^{37} atoms of ^{37}Cl. (The natural abundance of ^{37}Cl is 25%.) This is placed at the bottom of a 1-mile-deep mine in South Dakota to provide shielding against cosmic rays. The ^{37}Ar produced in the experiment is extracted, and identified by its characteristic β-decay. There is at present an unresolved and perplexing discrepancy between the counting rate expected from the standard solar model and the observed counting rate. The latter (0.4 ± 0.06 ^{37}Ar atoms per day) is too low by a factor of about four.

Other possible neutrino detector systems have been proposed to elucidate the problem, but they lack the cheapness and chemical simplicity of the ^{37}Cl experiment. The most popular suggestion would require some 40 tons of gallium, and the detection of germanium produced in the reaction

$$\nu + {}^{71}_{31}Ga + 0.23 \text{ MeV} \rightarrow {}^{71}_{32}Ge + e.$$

This experiment would be sensitive to the neutrinos produced in the basic p–p reaction (10.1) and disagreement with the theoretical prediction would be even more disconcerting than the discrepancy in the ^{37}Cl experiment.

10.5 Fusion reactors

For the generation of nuclear fusion power on Earth the immeasurably slow p–p reaction is useless. However, Coulomb barriers for the deuteron, $^{2}_{1}$H, are the same as for the proton, and the exothermic reactions

$$\begin{aligned} {}^{2}_{1}\mathrm{H}+{}^{2}_{1}\mathrm{H} &\rightarrow {}^{3}_{2}\mathrm{He}+\mathrm{n}+3.27\ \mathrm{MeV}, \\ {}^{2}_{1}\mathrm{H}+{}^{2}_{1}\mathrm{H} &\rightarrow {}^{3}_{1}\mathrm{H}+\mathrm{p}+4.03\ \mathrm{MeV}, \end{aligned} \qquad (10.9)$$

suggest deuterium to be a suitable fuel for a fusion power station. The natural abundance of deuterium is large, 0.015% of all hydrogen, and supplies of deuterium, in sea water for example, are effectively unlimited. The mass ratio of 2:1 makes isotope separation relatively easy.

Current research is more concerned with deuterium–tritium mixtures as fuel, using the reaction

$${}^{2}_{1}\mathrm{H}+{}^{3}_{1}\mathrm{H} \rightarrow {}^{4}_{2}\mathrm{He}+\mathrm{n}+17.62\ \mathrm{MeV}. \qquad (10.10)$$

This has two advantages over the reactions (10.9). First, the heat of reaction is greater. Second, and more important, the cross-section is considerably larger (Fig. 10.4), because of an excited state of $^{5}_{2}$He which gives a resonance in the cross-section. The principal disadvantage is that tritium, $^{3}_{1}$H, must be manufactured; it has no natural abundance since it undergoes β-decay with a mean life of 17.7 years. As Fig. 10.4 shows, the peak of $\overline{v\sigma}$ is at $kT = 60$ keV, and a temperature of 20 keV is regarded as a practical working temperature by fusion researchers.

A plasma at a temperature of 20 keV will vaporise any material container with which it comes into contact; current projects generally involve pulsed devices which contain and heat the plasma for short bursts of time only. For example, the moving electrically-charged particles of the plasma may be confined for short times, and even compressed, by magnetic fields, and heated by electromagnetic fields. Instruments such as the Joint European Torus (JET) at Culham are investigating these possibilities.

Inertial confinement, by the implosion of small pellets containing the deuterium–tritium fuel mixture, with the energy for implosion provided by pulsed laser beams, is another active area of research. A continuing series of 'mini-explosions' of such pellets, each containing a few milligrams of fuel, is envisaged. The scenario for such a reactor usually includes lithium in the heat-exchange blanket, since this provides a way of breeding tritium through the reactions

$$^{7}\text{Li}+\text{n}+2.46\ \text{MeV} \rightarrow {}^{3}\text{H}+\alpha+\text{n},$$

$$^{6}\text{Li}+\text{n} \rightarrow {}^{3}\text{H}+\alpha+4.8\ \text{MeV}.$$

(The natural abundances of ^{6}Li and ^{7}Li are 7.4% and 92.6% respectively.) The endothermic first reaction can be brought about by the fast neutrons produced in the deuterium–tritium reaction, and it is clear that in principle a breeding ratio of greater than one is possible.

To achieve a temperature T in a deuterium–tritium plasma there must be an energy input to the plasma of $4\rho_d(3k_BT/2)$ per unit volume, where ρ_d is the number density of deuterium ions and of tritium ions (i.e., $\rho_t=\rho_d$, and the electron density is $2\rho_d$, giving $4\rho_d$ particles per unit volume). The reaction rate in the plasma is $\rho_d^2\overline{\sigma v}$. If the plasma is confined for a time t_c then, per unit volume of plasma,

$$\frac{\text{fusion energy output}}{\text{energy input}}=\frac{\rho_d^2\overline{\sigma v}t_c(17.6\ \text{MeV})}{6\rho_d k_B T} \approx (10^{-19}\ \text{m}^3\ \text{s}^{-1})\rho_d t_c, \tag{10.11}$$

evaluating the right-hand side at $k_BT=20$ keV with the help of Fig. 10.4.

The plasma heating is certainly inefficient, so that a substantial fraction of useful energy is lost in this process, and the conversion of fusion energy to

10.4 Values of $\langle v\sigma\rangle$ (see equation (10.7)) for the combined deuterium–deuterium reactions (10.9) and the deuterium–tritium reaction (10.10). (Data from Keefe, D. (1982), *Ann. Rev. Nucl. Part. Sci.* **32**, 391.)

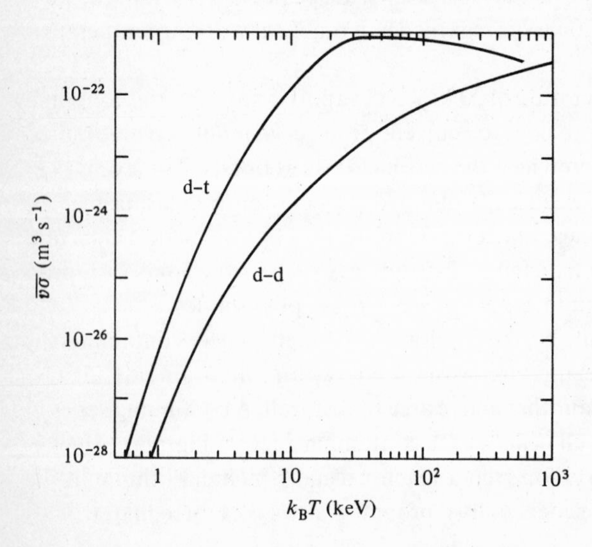

electricity is also (necessarily) inefficient. Hence a requirement for a useful device is that (fusion energy output)/(energy input) > 1, say. From equation (10.11), this is equivalent to the criterion $\rho_d t_c > 10^{19}\ \mathrm{m}^{-3}$ s. This is known as the *Lawson criterion*. More stringent formulations can be constructed for particular devices.

It should be appreciated that the engineering problems associated with either magnetic or inertial confinement as a basis for a working power station are immense and have not so far been solved in practice. The Lawson criterion provides an estimate of how close a particular design is to achieving practical results.

Problems

10.1(*a*) Assuming that the entire energy output from the Sun is derived from the PPI chain, estimate the flux of neutrinos at the Earth. (Distance of Earth from Sun $\sim 1.5 \times 10^8$ km.)

(*b*) The cross-section for a solar neutrino to interact with a nucleus is $\sim 10^{-20}$ b (cf. Problem 12.4). Show that such a neutrino incident on the Earth is very likely to pass through it unimpeded.

10.2 Why is the hydrogen content of the Earth so much less than that indicated in Table 10.1?

10.3 In the β-decay of ^{8}B, the neutrino takes on average about half of the energy released. Estimate the contribution to the Sun's luminosity per hydrogen atom consumed in the PPIII chain.

10.4 From Fig. 10.1, at the centre of the Sun $k_B T = 1.35$ keV and the mass density of hydrogen is $5.6 \times 10^4\ \mathrm{kg\ m}^{-3}$.

(*a*) Using equations (10.5), (10.6) and (10.8) estimate the contribution to the power density ε from the PPI chain. Compare your result with Fig. 10.1(*b*).

(*b*) For the ^{12}C–p reaction of the CNO cycle, $S(0) = 1.4$ keV b. Estimate the mean time it takes for a ^{12}C nucleus at the centre of the Sun to be converted to ^{13}N.

10.5 A deuterium–tritium plasma contains ρ_d deuterium and ρ_d tritium nuclei per unit volume.

(*a*) Show that to a good approximation $\rho_d(t)$ varies with time as $d\rho_d/dt = -\rho_d^2 \overline{\sigma v}$, where σ is the cross-section for the reaction

$\mathrm{d + t \rightarrow He + n} + 17.62$ MeV.

(*b*) If the plasma is brought together at time $t = 0$ with $\rho_d = \rho_0$, and confined for a time t_c at constant temperature, show that the proportion of the plasma 'burnt up' is

$$\frac{\overline{\sigma v}(\rho_0 t_c)}{1+\overline{\sigma v}(\rho_0 t_c)}.$$

(*c*) At $k_B T = 20\,\text{keV}$, $\overline{\sigma v} = 5\times10^{-22}\,\text{m}^3\,\text{s}^{-1}$. What Lawson number $(\rho_0 t_c)$ would be required to burn 5% of the fuel?

11
Nucleosynthesis in stars

In the preceding chapter we explained how in a star like the Sun helium is steadily formed from the fusion of hydrogen. In this chapter we sketch some of the basic ideas of 'nuclear astrophysics', a subject which seeks to understand all the nuclear processes leading to energy generation in stars in the various stages of stellar evolution, and to account for the observed relative abundances of the elements in the Solar System in terms of these processes.

The accepted theory of the Universe is that it is expanding, and began with an intensely hot and dense 'big bang' between 10×10^9 and 20×10^9 years ago. A few hundred thousand years after the big bang, the expanding material had cooled sufficiently for it to condense into a gas made up of hydrogen and helium atoms in a ratio of about 100:7 by number, together with photons and neutrinos. Apart from a small amount of lithium, it is thought that the proportion of heavier elements produced in this first explosion was insignificant (essentially because there are no stable nuclei with $A=5$ or $A=8$). If this is so, we must conclude that all the heavier nuclei in the Solar System have been produced in previous generations of stars and then thrown out into space again, perhaps in the explosion of supernovae.

11.1 Stellar evolution

Consider a star which has condensed from the primordial

hydrogen–helium mixture, and in which hydrogen burning has set in at the core. As the hydrogen in the core is consumed, the reaction rate eventually becomes insufficient to sustain the temperature, and hence the pressure, that prevents further gravitational contraction. Thus more material falls into the core region. If the star is massive enough, the gravitational energy released raises the temperature of the core sufficiently for helium to begin burning at a significant rate. As the helium in turn is consumed, further stages of nuclear burning set in until the most tightly-bound elements, iron and nickel, are formed. At each stage a higher temperature is needed to overcome a higher Coulomb barrier; the energy for this is provided by gravitational contraction.

Before considering these later stages of nuclear burning in more detail, it is important to appreciate that there are conditions under which the central pressure can permanently balance the pressure exerted by gravity. Then contraction will cease and the temperatures for further steps in nucleosynthesis will not be reached. After completing as much burning as it can, the star will simply cool. The first contribution to the pressure that may stop contraction is the 'electron-degeneracy pressure'. Since electrons are fermions, it follows from the Pauli principle that, even in a cold star with $T \sim 0$ K, electron states are occupied up to an energy $\varepsilon_F = (\hbar^2/2m_e)k_F^2$, where (Appendix B, equation (B.5)) $k_F^3 = 3\pi^2\rho_e$, and ρ_e is the number density of electrons. Thus in matter at high density there exist electrons with high k_F and hence high kinetic energy, which necessarily exert a high pressure. To obtain a simple order-of-magnitude estimate of this effect, we set the density of matter in a star of mass M, radius R, to be constant. Then the number of electrons in the star is $N_e = (M/\mu)$, where μ is the stellar mass per electron. For material with $Z = N$, we have $\mu = 2$ amu, to a good approximation. The electron density is $\rho_e = (M/\mu)/(4\pi R^3/3)$, giving

$$k_F^3 = 3\pi^2\rho_e = \left(\frac{9\pi}{4}\right)\left(\frac{M}{\mu}\right)\frac{1}{R^3}. \tag{11.1}$$

Assuming that the electrons can be treated non-relativistically, the total kinetic energy of N_e electrons at $T = 0$ is $(3/5)N_e\varepsilon_F$ (cf. Problem 5.2). At $T \sim 0$ K, the sum of the electron kinetic energies and the gravitational potential energy is therefore

$$E = \frac{3}{5}\left(\frac{M}{\mu}\right)\frac{\hbar^2}{2m_e}\left(\frac{9\pi M}{4\mu}\right)^{\frac{2}{3}}\frac{1}{R^2} - \frac{3}{5}\frac{GM^2}{R}. \tag{11.2}$$

The star begins its life with R large, and the electron energy is then much smaller than the gravitational energy. As the star evolves it contracts, so that our 'model' R decreases and E becomes more negative. The energy released goes into heating the interior of the star and into radiation. However, no more energy can be released by contraction when E reaches its minimum value where $\mathrm{d}E/\mathrm{d}R=0$, at

$$R_{\min}=\left(\frac{9\pi}{4}\right)^{\frac{2}{3}}\frac{\hbar^2}{Gm_e M^{\frac{1}{3}}\mu^{\frac{5}{3}}}$$

$$=7.2\left(\frac{M_\odot}{M}\right)^{\frac{1}{3}}\times 10^3\ \mathrm{km}. \qquad (11.3)$$

(A calculation which does not make our assumption of constant density, but determines the density self-consistently, gives a numerical coefficient of 8.8 instead of 7.2.)

The corresponding mass density in our model at this minimum radius is

$$\rho_{\text{mass}}=\frac{M}{(4\pi R_{\min}^3/3)}=\frac{4M^2G^3m_e^3\mu^5}{27\pi^3\hbar^6}$$
$$=1.27\left(\frac{M}{M_\odot}\right)^2\times 10^9\ \mathrm{kg\ m^{-3}}. \qquad (11.4)$$

There are many stars with masses similar to, but generally smaller than, the Sun which are close to this inert condition. They have high density and small radii, and are called white dwarfs. It may be noted that the minimum radius decreases as the mass increases.

The maximum electron momentum in our model when $R=R_{\min}$ is, using equation (11.1),

$$p_F=\hbar k_F=(3\pi^2)^{\frac{1}{3}}(\rho_{\text{mass}}/\mu)^{\frac{1}{3}}\hbar=0.44\left(\frac{M}{M_\odot}\right)^{\frac{2}{3}}\ \mathrm{MeV}/c.$$

Since the rest mass of an electron is $0.511\ \mathrm{MeV}/c^2$, the assumption in our calculation above that the electron can be treated non-relativistically, i.e. that $p_F \ll m_e c$, is clearly suspect for stars with $M\approx M_\odot$ and is certainly wrong for stars of large M. In the limit when M is large, we take

$$\varepsilon=(p^2c^2+m_e^2c^4)^{\frac{1}{2}}\approx pc=\hbar ck,$$

for all the electrons, so that the total energy of N_e electrons at $T=0$ becomes $(3/4)N_e(\hbar ck_F)$ (Problem 11.1). Hence, using equation (11.1) again, the expression (11.2) for the energy is replaced by

$$E\approx\left[\frac{3}{4}\hbar c\left(\frac{9\pi}{4}\right)^{\frac{1}{3}}\left(\frac{M}{\mu}\right)^{\frac{4}{3}}-\frac{3}{5}GM^2\right]\frac{1}{R}. \qquad (11.5)$$

If M is sufficiently large the coefficient of $(1/R)$ is negative and there is no minimum energy: electron degeneracy alone cannot prevent the collapse of the star. Our extreme relativistic approximation becomes increasingly valid as R decreases. Equation (11.5) suggests that the critical value of M is

$$M = \frac{15}{16} \frac{(5\pi)^{\frac{1}{2}}}{\mu^2} \left(\frac{\hbar c}{G}\right)^{\frac{3}{2}} = 1.74\, M_{\odot}.$$

A more careful calculation takes proper account of relativistic energies and determines the density distribution self-consistently. It is then found that the electron degeneracy pressure cannot stop the gravitational collapse of a star of mass M if $M > 1.44\, M_{\odot}$. This result, due to Chandrasekhar, is known as the *Chandrasekhar limit*.

At very high densities of matter it becomes energetically favourable for electrons to be captured by protons, and a Fermi gas of neutrons is formed. Thus final collapse may be prevented by neutron-degeneracy pressure. The mass density of what is then a *neutron star* is very high, comparable with the density of nuclear matter, as equation (11.4) indicates if we replace m_e by m_n and put $\mu = 1$. Under these conditions our simple expressions which neglect nuclear interactions are of limited validity!

11.2 From helium to silicon

We return now to the problem of nucleosynthesis beyond helium. It is clear that the fusion of hydrogen to helium already converts most of the available nuclear potential energy into heat and radiation. The binding energy per nucleon in ^{4}He is 7.1 MeV, and there is only a further 1.7 MeV per nucleon to be released in complete burning to iron. Also, as can be seen from Fig. 10.3, as the elements involved become heavier and more charged, higher and higher temperatures are required for there to be significant tunnelling through the Coulomb barriers. In fact, as we shall see, the simple fusion process is superseded by another when elements around $^{28}_{14}$Si have been produced.

A few of the important reactions associated with helium burning to oxygen, and oxygen to silicon, are listed below, along with typical temperatures and mass densities at which in a sufficiently massive star they are calculated to occur:

$$\left.\begin{aligned} &^4\text{He} + {}^4\text{He} \to {}^8\text{Be} \\ &^4\text{He} + {}^8\text{Be} \to {}^{12}\text{C} + \gamma + 12 \times (0.61\ \text{MeV}) \\ &^4\text{He} + {}^{12}\text{C} \to {}^{16}\text{O} + \gamma + 16 \times (0.45\ \text{MeV}) \end{aligned}\right\} \quad \begin{aligned} &k_B T \sim (10\text{–}20\ \text{keV}) \\ & \\ &\rho \sim (10^5\text{–}10^8)\ \text{kg m}^{-3} \end{aligned}$$

$$^{16}\text{O} + {}^{16}\text{O} \to {}^{28}_{14}\text{Si} + {}^4\text{He} + 32 \times (0.30\ \text{MeV}) \quad k_B T \sim (100\text{–}200)\ \text{keV}$$

$$\rho \sim 10^9\ \text{kg m}^{-3}.$$

The initial stage of helium burning needs some explanation. As Table 4.2 indicated, ^{4}He has the largest binding energy per nucleon of any nucleus less massive than ^{12}C. The most stable form of nuclear material with $A < 12$ is therefore ^{4}He, and in particular $^8_4$Be does not exist as a stable nucleus. Nevertheless $^8_4$Be exists as a resonant state that is seen in the laboratory in α–α scattering at an energy of 94 keV, in the centre-of-mass frame, with a narrow width (due to the Coulomb barrier) of 2.5 eV. In a ^{4}He plasma this state is established with an equilibrium density such that the rate of production equals the rate of decay. Thus the 'mass gaps' at $A = 5$ and $A = 8$ can be bridged. The next step in the chain, $^8\text{Be} + \alpha \to {}^{12}\text{C} + \gamma$, is in fact enhanced because it is a resonant reaction. There is an excited state of ^{12}C at 0.29 MeV above the $^8\text{Be} + \alpha$ threshold.

In the final stages of oxygen burning, core temperatures in the star are calculated to reach 300–400 keV, with mass densities in excess of 10^9 kg m^{-3}.

11.3 Silicon burning

In all of the preceding stages of stellar evolution, photons have always been present in thermal equilibrium with the plasma. They have played an important role in radiative heat transfer, but have been unimportant for initiating the nuclear processes we have discussed. However, a photon couples electromagnetically to a nucleus and can be readily absorbed by a nucleus to form an excited state. If the photon has an energy above the threshold for nuclear break-up of that nucleus, break-up can occur. This process is called *photodisintegration.*

As the temperature in the core of a star approaches $k_B T = 1$ MeV, the increasing number of photons in the high-energy tail of the thermal distribution makes photodisintegration an important process. In particular, protons, neutrons and α-particles are knocked out of nuclei. Although this effectively undoes some of the nuclear building that has gone on before, protons and α-particles, as well as neutrons, are at these temperatures readily accepted into any nucleus present, and a situation approaching thermal equilibrium is quickly established with the most tightly bound elements, iron and nickel, copiously produced.

At this stage the core of a massive star is in an unstable condition. There is no more nuclear fuel to burn to delay further gravitational contraction, so even higher densities and temperatures occur. It then becomes energetically advantageous for electrons at the top of the Fermi distribution to undergo electron capture to form neutron-rich nuclei, which on Earth would be β^--unstable. This process removes heat from the core by producing neutrinos which escape, as well as removing electrons. Thus the pressure falls, hastening contraction and leading to the removal of even more electrons.

Eventually there will be a catastrophic collapse of the core, an implosion which can only be stopped by nucleon pressure and the nucleon–nucleon short-range repulsion.

The cooler regions of the star outside the core will contain unburned or only partially-burned material. As the core implodes, these regions will quickly fall inwards and rise in temperature so that the remaining fuel burns explosively, blowing the stellar outer mantle into space. This is the scenario for neutron star formation accompanied by a *supernova explosion*.

11.4 Nucleosynthesis of heavy elements

The most likely process for the formation of elements heavier than those grouped around iron, produced in the silicon burning described above, is neutron capture. If a supply of free neutrons is available, they can accrete on an iron-group seed nucleus by radiative capture, unimpeded by Coulomb barriers, to build-up a neutron-rich isotope. As the neutron number in the nucleus increases it will become unstable to β^--decay, thus forming a new element of atomic number $Z+1$ from an element of atomic number Z. Successive neutron captures, interspersed with β^--decays, can eventually build up many, but not all, of the heavy stable nuclei. Since the build-up follows the neutron-rich side of the 'β stability valley' (§ 4.6), some of the proton-rich stable isotopes are inaccessible in this process. It is an interesting fact that such isotopes have a much smaller natural abundance than their neutron-rich neighbours.

There are two basic time scales in this scenario of heavy element synthesis by neutron accretion: the β-decay lifetimes and the time intervals between successive neutron captures (which are inversely proportional to the capture cross-sections and the neutron flux). If the rate of neutron capture is slow compared with the relevant β-decay rates (the s-process) the nuclei that are built up will follow the bottom of the β stability valley very closely. If the rate of neutron capture is rapid (the r-process) highly unstable neutron-rich isotopes will be formed which cascade down to stable nuclei, some of which are inaccessible by the s-process; thorium and uranium must have been formed in this way. The observed nuclear abundances, especially in the regions of closed-shell nuclei, suggest that both the r- and the s-processes have played a part in the synthesis of nuclei found in the Solar System and, in particular, the heavy elements found on Earth.

The site of the r-process is believed to be in supernovae explosions close to the region of neutron star formation, where over a short period of time large neutron fluxes can be expected. The s-process probably occurs during helium burning in massive stars, where a low neutron flux can be provided by a number of reactions, for example (Fig. 8.4),

$$\alpha + {}^{13}_{6}\mathrm{C} \rightarrow {}^{16}_{8}\mathrm{O} + \mathrm{n}.$$

For nucleosynthesis of the heavy elements by the s-process there must be iron present, derived from nucleosynthesis in previous generations of stars and forming part of the gas from which the star in question condensed.

In this chapter we have attempted to provide no more than a qualitative sketch of a theory which is still being developed. Many of the basic components of the theory are probably in place but important aspects are still being investigated through laboratory measurements and theoretical estimates of reaction rates, and computer studies of reaction networks, combined with stellar models. A rich variety of facts and phenomena remain to be explained.

Problems

11.1 In a plasma with high electron number density ρ_e, using the extreme relativistic approximation in which energy and momentum are related by $E = pc$, show that the average energy of an electron is $(3/4)\hbar c k_F$, where k_F is given by $k_F^3 = 3\pi^2 \rho_e$.

11.2 The planet Jupiter is composed mostly of hydrogen. It has mass 1.9×10^{27} kg and mean radius $\approx 7 \times 10^7$ m. Show that if it were uniformly dense its gravitational energy per particle would be only 7 eV, too small to ignite nuclear reactions.

11.3 Estimate the mass density of (metallic) hydrogen at 0 K at which it is energetically favourable to subtract electrons from the electron gas and form neutrons by the inverse β-decay

$$\mathrm{p} + \mathrm{e}^- \rightarrow \mathrm{n} + \nu.$$

11.4 If the Sun were a neutron star, what would be its radius and its mass density?

11.5 The Planck radiation law states that the number of photons per unit volume in an energy range $\mathrm{d}E$ is

$$\frac{1}{\pi^2(\hbar c)^3} \frac{E^2\,\mathrm{d}E}{e^{E/k_B T} - 1}.$$

At a temperature $k_B T = 500$ keV, estimate the number of photons per unit volume with an energy greater than 8 MeV.

11.6 The cross-section for ^{8}Be production in α–α scattering at energy E in the centre-of-mass frame is given by the Breit–Wigner formula

$$\sigma(E) = \frac{2\pi\hbar^2}{m_\alpha E} \frac{\Gamma^2}{(E - E_0)^2 + \Gamma^2/4},$$

with $E_0=94$ keV, $\Gamma=2.5$ eV. (Note the additional factor of 2 in the Breit–Wigner formula for identical particles: see Problem D.1.) In a plasma at temperatures $k_B T \gg \Gamma$, the thermal average $\overline{v\sigma}$ is dominated by energies in the neighbourhood of E_0.

(*a*) Show that

$$\overline{v\sigma} \approx 16\hbar^2\Gamma\left(\frac{\pi}{m_\alpha k_B T}\right)^{\frac{3}{2}} e^{-E_0/k_B T}.$$

(*b*) Hence show that the density ρ_{Be} of ^{8}Be in equilibrium with α-particles of number density ρ_α is in the ratio

$$\frac{\rho_{Be}}{\rho_\alpha}=\frac{\rho_\alpha}{2}\left(\frac{\hbar}{\Gamma}\right)\overline{v\sigma}=\rho_\alpha\left(\frac{4\pi\hbar^2}{m_\alpha k_B T}\right)^{\frac{3}{2}} e^{-E_0/k_B T}.$$

(*c*) Calculate this ratio for $k_B T=15$ keV and a helium mass density of 10^6 kg m^{-3}.

12
Beta-decay and gamma-decay

In this chapter we present some of the theory of β-decay and γ-decay. In both cases, a fuller treatment requires more quantum mechanics than is usually contained in an undergraduate course, but we shall see that much of the experimental phenomena can be understood qualitatively without the complete relativistic theory.

12.1 What must a theory of β-decay explain?

In β-decay, introduced in § 3.5, the charge of a nucleus changes while A remains fixed. This occurs either by the simultaneous emission of an electron and an anti-neutrino, or a positron and a neutrino, or by the capture of an atomic electron with the emission of a neutrino. The appropriate stability conditions were discussed in § 4.6. Several nuclei, for example ^{64}Cu, can decay by any of these processes (Fig. 4.5). In electron capture, the neutrino energy and the recoil energy of the nucleus are sharply defined. In the other processes, the electron (or positron) can take any energy between zero and the maximum allowed by energy conservation. Figure 12.1 shows the experimentally-determined energy spectra for electron emission and for positron emission from $^{64}_{29}$Cu. It was the observation of continuous energy distributions such as these that led Pauli to infer the existence of the neutrino in 1933: given that the energy levels of a nucleus are discrete, the electron and nuclear recoil energies in the centre-of-mass frame would by energy and momentum conservation be likewise discrete, unless some third particle (the neutrino) were present to share energy and momentum. The neutrino mass can be deduced to be small since, when the electron takes its maximum energy, the energy balance to within the accuracy of present experiments is complete. Maximum electron energy corresponds to the neutrino carrying no momentum, so that the neutrino energy then would be its rest-mass energy $m_\nu c^2$. We consider recent experimental limits on m_ν in § 12.5. Because the neutrino interacts so

12.1 Electron and positron energy spectra from the β-decay of ^{64}Cu, giving the probability distributions in energy (both normalised to unity) from a large sample of decays. The experimental points are from Langer, L. M. *et al.* (1949), *Phys. Rev.* **76**, 1725. The curves are fits to the data using equations (12.5) and (12.6).

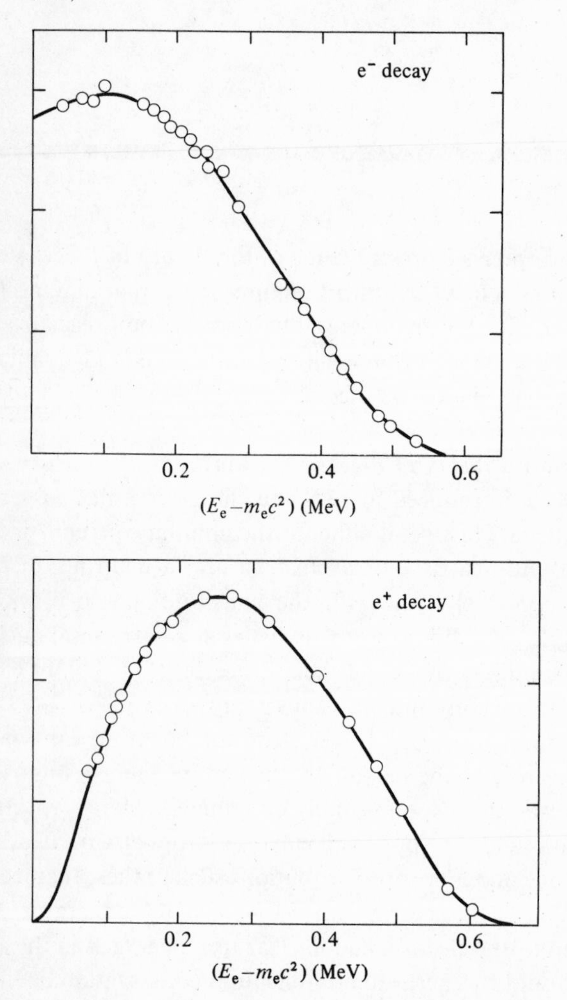

weakly with other particles, it was not until 1959 that its existence was more directly confirmed by the observation of the reaction

$$\bar{\nu}_e + p \rightarrow n + e^+,$$

using the high neutrino flux associated with a nuclear reactor (which arises from β-decays of the neutron-rich fission fragments).

As in the case of γ-decay mean lives, β-decay mean lives span many orders of magnitude. For example, the most common isotope of indium, $^{115}_{49}\mathrm{In}$, is β-unstable, but its mean life is $\sim 10^{14}$ years, whereas β-decay mean lives of the order of seconds or minutes are common. As with γ-decay, mean lives depend strongly on the nuclear-spin change in the decay.

The first experimental evidence for the violation of mirror symmetry at the subatomic level was found in β-decay by Wu in 1957, following a suggestion by Lee and Yang. The experiment measured the angular distributions of electrons from the decay of $^{60}\mathrm{Co}$:

$$^{60}_{27}\mathrm{Co} \rightarrow {}^{60}_{28}\mathrm{Ni} + e^- + \bar{\nu}.$$

The $^{60}\mathrm{Co}$ nucleus, of spin $5\hbar$, has a large magnetic moment, and the nuclei in the sample were polarised by a magnetic field. The electrons from the decays were observed to be preferentially emitted in the opposite direction to the nuclear spins. Such a correlation violates the principle of mirror symmetry, since in the mirror image of this experiment, more electrons appear to be emitted in the same direction as the nuclear spin. An examination of Fig. 12.2 will make this clear. (*Any* mirror plane may be chosen, and will lead to the same conclusion.) Since the description of the experiment and of its mirror image differ, it follows that parity cannot be a symmetry of the weak interaction (§ 2.6). The breakdown of parity conservation in the decay of the muon, described in § 2.6, was discovered shortly afterwards.

12.2 A schematic representation of the $^{60}\mathrm{Co}$ decay experiment in real space (*a*), and the mirror image of the experiment (*b*). In both (*a*) and (*b*) the spin of the cobalt nucleus is pointing to the right; the spin is a pseudo-vector which does not change direction under this reflection. The sample is polarised by the magnetic field produced by a current flowing in the direction indicated.

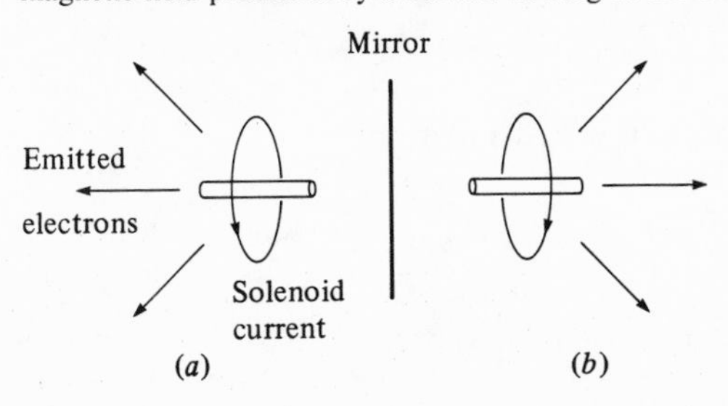

12.2 The Fermi theory of β-decay

A simple theory of β-decay was suggested by Fermi in 1934. Although this theory is incomplete (it does not allow for parity violation, for example), it is able to describe the spectra of Fig. 12.1, and gives a qualitative understanding of the range of β-decay mean lives.

To be specific, we consider in the shell model a decay in which a proton outside a doubly-closed shell changes to a neutron:

$$^{17}_{9}\mathrm{F} \to {}^{17}_{8}\mathrm{O} + \mathrm{e}^{+} + \nu.$$

A proton in the closed shell cannot change into a neutron since the neutron shell is also full and the Pauli principle forbids the transition. Thus the nucleons in the closed shells play no part in the decay and we can take the initial state of the system to be simply

$$\Psi_0 = \psi_{\mathrm{p}}(\mathbf{r}_{\mathrm{p}}),$$

where ψ_{p} is the state of the single proton in the $\mathrm{d}_{\frac{5}{2}}$ shell. The final state of the system consists of a neutron in the same shell, a positron e^{+}, and a neutrino ν

$$\Psi_{\mathrm{f}} = \psi_{\mathrm{n}}(\mathbf{r}_{\mathrm{n}})\psi_{\mathrm{e}}(\mathbf{r}_{\mathrm{e}})\psi_{\nu}(\mathbf{r}_{\nu}),$$

(in an obvious notation). Note that we are ignoring the spins of the particles involved and neglecting the recoil of the nucleus.

The transition rate from Ψ_0 to Ψ_{f} is given in perturbation theory by:

$$\text{transition rate} = \frac{2\pi}{\hbar}|H_{\mathrm{f}0}|^2 n_{\mathrm{f}}(E_0), \tag{12.1}$$

where $H_{\mathrm{f}0}$ is the matrix element linking the initial and final states, and $n_{\mathrm{f}}(E_0)$ is the density of (specified) states Ψ_{f} at the energy E_0 released in the decay. This result is obtained in Appendix D (equation (D.6)) and is often called 'Fermi's golden rule' in texts on quantum mechanics.

We saw in Chapter 2 that the weak interaction responsible for β-decay, mediated by the heavy bosons $\mathrm{W}^{\pm}$, is of very short range $\approx 2\times 10^{-3}$ fm. This fact was anticipated by Fermi, who suggested that at the moment of interaction all particles were at the same point in space, so that the interaction matrix element

$$H_{\mathrm{f}0} = \int \Psi_{\mathrm{f}}^{*} H \Psi_0 \, \mathrm{d}^3\mathbf{r}_{\mathrm{n}} \, \mathrm{d}^3\mathbf{r}_{\mathrm{p}} \, \mathrm{d}^3\mathbf{r}_{\mathrm{e}} \, \mathrm{d}^3\mathbf{r}_{\nu}$$

could be of the form

$$H_{\mathrm{f}0} = G_{\mathrm{F}} \int \psi_{\mathrm{n}}^{*}(\mathbf{r})\psi_{\mathrm{e}}^{*}(\mathbf{r})\psi_{\nu}^{*}(\mathbf{r})\psi_{\mathrm{p}}(\mathbf{r}) \, \mathrm{d}^3\mathbf{r},$$

where the *Fermi constant* G_F is a measure of the strength of the weak interaction.

We may take for the neutrino a plane-wave state

$$\psi_\nu(\mathbf{r}) = \frac{1}{V^{\frac{1}{2}}} e^{i\mathbf{k}_\nu \cdot \mathbf{r}_\nu},$$

with the wave-function normalised in an arbitrarily large volume V for mathematical convenience. We take the positron wave-function to be also a plane wave

$$\psi_e(r) = \frac{1}{V^{\frac{1}{2}}} e^{i\mathbf{k}_e \cdot \mathbf{r}_e},$$

though this is only a first approximation; since a positron (or electron) is charged, its wave-function will be modified by the Coulomb field of the daughter nucleus. The matrix element H_{f0} becomes in this approximation

$$H_{f0} = \frac{G_F}{V} \int \psi_n^*(\mathbf{r}) \psi_p(\mathbf{r}) e^{-i(\mathbf{k}_e + \mathbf{k}_\nu) \cdot \mathbf{r}} \, d^3\mathbf{r}.$$

The energies involved in β-decay are generally at most a few MeV, and the corresponding momenta $\hbar\mathbf{k}_e$, $\hbar\mathbf{k}_\nu$ a few MeV/c. Hence the wave vectors $\mathbf{k}_e$, $\mathbf{k}_\nu$ are $\sim$MeV/$\hbar c \sim 10^{-2}$ fm, and the exponent of the exponential in the integral is small over the range of the nuclear wave-functions. It is therefore an excellent approximation to expand the exponential to give

$$H_{f0} = \frac{G_F}{V} \int \psi_n^*(\mathbf{r}) \psi_p(\mathbf{r}) \, d^3\mathbf{r} - \frac{iG_F}{V} (\mathbf{k}_e + \mathbf{k}_\nu) \cdot \int \psi_n^*(\mathbf{r}) \psi_p(\mathbf{r}) \mathbf{r} \, d^3\mathbf{r} + \cdots,$$

and keep only the first non-vanishing term. This argument is clearly also valid when the nuclear wave-functions involved are more complicated than in our simple example.

A decay is said to be allowed if the first term is finite. It is said to be first forbidden if the first term is zero as happens for example if the initial and final nuclear states are of opposite parity, but not the second, and so on. The diversity of β-decay rates is largely accounted for by the degree of forbiddenness of the transition and this in turn by the change in nuclear spin (as in γ-decay). In the case of indium, previously cited, the first term in the expansion not to vanish is found to be the fifth and the decay is fourfold forbidden.

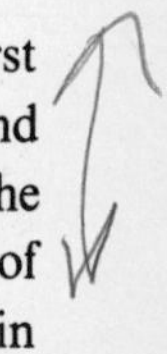

We shall concentrate our discussion on allowed transitions, in which case the matrix element is

$$H_{f0} = (G_F/V)M_F, \tag{12.2}$$

where M_F is the appropriate nuclear matrix element. In our example of the decay of ^{19}F, the spatial shell model wave-functions of the proton and neutron are the same apart from Coulomb effects, and hence

$$M_F = \int \psi_n^*(\mathbf{r})\psi_p(\mathbf{r})\, d^3\mathbf{r} \approx 1.$$

12.3 Electron and positron energy spectra

Consider an allowed β-decay in which the electron is emitted in a particular state of (relativistic) energy E_e. For simplicity we neglect the recoil energy of the daughter nucleus, which is in any case always a small correction. Then the relativistic neutrino energy E_ν is given by $E_0 = E_e + E_\nu$, where E_0/c^2 is the nuclear mass difference. The density of final states factor in the formula (12.1) is thus the density of neutrino states, $n_\nu(E_0 - E_e)$, at energy $(E_0 - E_e)$. There are $n_e(E_e)\,dE_e$ electron states with energies between E_e, $E_e + dE_e$, where n_e is the density of electron states. Thus the total transition rate dR for decays to electron states with energies in the range E_e, $E_e + dE_e$, is

$$dR = \frac{2\pi}{\hbar} |H_{f0}|^2 n_\nu(E_0 - E_e) n_e(E_e)\, dE_e. \tag{12.3}$$

We need expressions for the densities of states n_ν, n_e. As is explained in Appendix B, neglecting spin there are $[V/(2\pi)^3]4\pi k^2\, dk$ plane-wave states with $|\mathbf{k}|$ in the range $k, k + dk$. For the relation between E and k we must use for both electrons and neutrinos the relativistic formula

$$E^2 = p^2c^2 + m^2c^4 = (\hbar k)^2c^2 + m^2c^4,$$

so that $E\, dE = \hbar^2c^2k\, dk$, $k = (E^2 - m^2c^4)^{\frac{1}{2}}/\hbar c$. Thus the relativistic density of states formula for a particle of mass m is

$$n(E)\, dE = \frac{V}{(2\pi)^3} \frac{4\pi}{\hbar^3c^3} (E^2 - m^2c^4)^{\frac{1}{2}} E\, dE. \tag{12.4}$$

Note that E is here the total energy, which includes the rest-mass energy. Substituting in equation (12.3) and using (12.2) for the matrix element of an allowed transition gives

$$dR = \frac{G_F^2|M_F|^2}{2\pi^3\hbar^7c^6} S_0(E_e)\, dE_e \tag{12.5}$$

where

$$S_0(E_e) = [(E_0 - E_e)^2 - m_\nu^2c^4]^{\frac{1}{2}}(E_0 - E_e)(E_e^2 - m_e^2c^4)^{\frac{1}{2}} E_e.$$

The arbitrary normalisation volume V cancels out from the final result (12.5), as we should expect. We have allowed the neutrino mass to be finite in order to discuss the experimental evidence for its small or zero value.

The electron (positron) energy dependence in the transition rate (12.5) comes *entirely* from the lepton density-of-state factors included in $S_0(E)$: the other factors are independent of electron energy. The formula can be improved by allowing for the interaction between the electron and the Coulomb field of the daughter nucleus of charge Z_d. Since only the electron wave-function at the nucleus is important, $S_0(E_e)$ is modified to

$$S_c(E_e) = F(Z_d, E_e)S_0(E_e), \tag{12.6}$$

where

$$F(Z_d, E_e) = \left|\frac{\psi_e(Z_d, 0)}{\psi_e(0, 0)}\right|^2,$$

and $\psi_e(Z_d, \mathbf{r})$ is the electron wave-function in the Coulomb potential $\pm Z_d e^2/4\pi\varepsilon_0 r$. Extensive tables of $F(Z, E)$ are available for precise calculations, but a simple approximation is the non-relativistic formula

$$F(Z, E_e) = \frac{2\pi\eta}{1 - e^{-2\pi\eta}},$$

where $\eta = \pm Ze^2/(4\pi\varepsilon_0 \hbar v)$; the positive sign holds for electrons, the negative for positrons, and v is the electron (positron) final velocity.

As $v \to 0$, $F(Z, E_e) \to 2\pi\eta$ for electrons. The $(1/v)$ factor makes $S_c(E_e)$ non-vanishing at the origin, where $E_e \to m_e c^2 + \frac{1}{2}m_e v^2$; the decay rate is enhanced at low energies since the Coulomb field for electrons is attractive.

For positrons at low energies $F(Z, E_e) \to 2\pi|\eta|e^{-2\pi|\eta|}$. The Coulomb field is repulsive for positrons and we can recognise the exponential as the tunnelling factor through the Coulomb barrier, which suppresses positron emission at low energies.

Figure 12.1 shows fits to the experimental statistical spectra for electron and positron emission from ^{64}Cu. The Coulomb-corrected $S_c(E_e)$ give excellent agreement with the shapes of the experimental curves. The more detailed theory of β-decay retains this factor.

The total transition rate for a particular allowed decay is obtained by integrating the partial rate (12.5) over all electron energies, to give for the mean lifetime τ the formula

$$\frac{1}{\tau} = \frac{G_F^2|M_F|^2 m_e^5 c^4}{2\pi^3\hbar^7} f(Z_d, E_0),$$

where

$$f(Z, E_0) = \left(\frac{1}{m_e c^2}\right)^5 \int_{mc^2}^{E_0} F(Z, E_e)(E_0 - E_e)^2(E_e^2 - m_e^2 c^4)^{\frac{1}{2}} E_e \, dE_e. \tag{12.7}$$

To obtain (12.7) we have set $m_\nu = 0$. $f(Z, E)$ is a dimensionless function for which again there are extensive tables. Some representative graphs are given in Fig. 12.3.

12.4 Electron capture

In an atomic environment, β-decay by electron capture always competes with positron emission, and is sometimes the only energetically allowed β-decay. To take our previous example, ${}^{17}_{9}\mathrm{F}$ can also decay by electron capture:

$${}^{17}_{9}\mathrm{F} + \mathrm{e}^- \to {}^{17}_{8}\mathrm{O} + \nu.$$

The electron and proton are now both in the initial state,

$$\Psi_0 = \psi_\mathrm{p}(\mathbf{r}_\mathrm{p})\psi_\mathrm{e}(\mathbf{r}_\mathrm{e}),$$

and the electron is most likely to be a K-shell electron, since K-shell wave-functions have the greatest overlap with the nucleus. To a good approximation, these wave functions are hydrogen-like, little influenced by the outer shell atomic electrons, so we can take

$$\psi_\mathrm{e}(\mathbf{r}) = \pi^{-\frac{1}{2}}\left(\frac{Zm_\mathrm{e}e^2}{4\pi\varepsilon_0\hbar^2}\right)^{\frac{3}{2}} \exp\left(-\frac{Zm_\mathrm{e}e^2 r}{4\pi\varepsilon_0\hbar^2}\right),$$

12.3 The function $f(Z, E_0)$. The sequence of curves is for $Z = 90$, 60, 30 and 0 for e^- decay and continuing with $Z = 30$, 60, 90 for e^+ decay. (Formulae can be found in Feenberg, E. & Trigg, G. (1950), *Rev. Mod. Phys.* **22**, 399.)

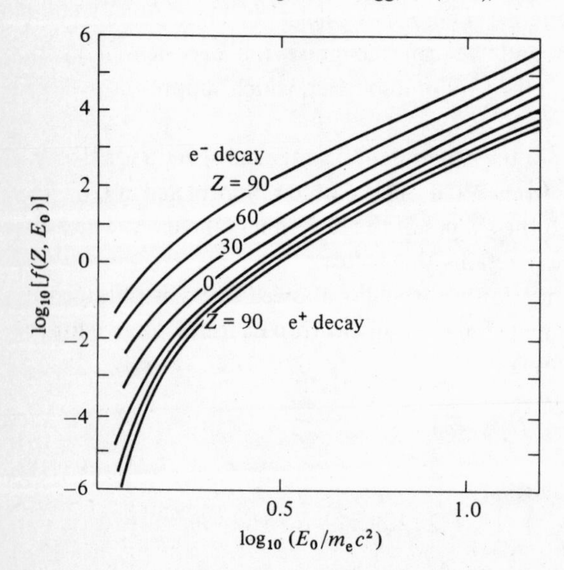

where Z is the atomic number of the parent nucleus. The final state is

$$\Psi_f = \psi_n(\mathbf{r}_n)\psi_\nu(\mathbf{r}_\nu).$$

For the neutrino we again take a plane-wave state $V^{-\frac{1}{2}}e^{i\mathbf{k}\cdot\mathbf{r}}$ normalised in a volume V.

In the simple Fermi theory, we now have

$$H_{f0} = G_F \int \psi_n^*(\mathbf{r})\psi_\nu^*(\mathbf{r})\psi_p(\mathbf{r})\psi_e(\mathbf{r})\,d^3\mathbf{r}.$$

Assuming that the transition is allowed, this reduces to

$$H_{f0} = \frac{G_F}{V^{\frac{1}{2}}}\psi_e(0)\int \psi_n^*(\mathbf{r})\psi_p(\mathbf{r})\,d^3\mathbf{r} = \frac{G_F}{V^{\frac{1}{2}}\pi^{\frac{1}{2}}}\left(\frac{Zm_e e^2}{4\pi\varepsilon_0\hbar^2}\right)^{\frac{3}{2}} M_F,$$

since the electron wave-function can be treated as constant over the nuclear volume.

Neglecting the nuclear recoil, the emitted neutrino has energy E_ν, where E_ν/c^2 is the *atomic* mass difference $\approx (E_0/c^2 + m_e)$ (cf. (4.13)). The appropriate density of states in the formula for the transition rate is the neutrino density of states at this energy, given by equation (12.4). Setting $m_\nu = 0$, we obtain

$$\text{decay rate for electron capture} = \frac{2\pi}{\hbar}|H_{f0}|^2 n_\nu(E_\nu)$$

$$= \frac{G_F^2|M_F|^2E_\nu^2}{\pi^2\hbar^4c^3}\left(\frac{Zm_e e^2}{4\pi\varepsilon_0\hbar^2}\right)^3.$$

Consistently with our neglect of electron spin, only one K electron is included in the calculation. The ratio of the electron capture rate R_K to the positron emission rate R_{e^+} is independent of G_F and the nuclear matrix element, and is

$$\frac{R_K}{R_{e^+}} = 2\pi\left(\frac{E_\nu}{m_e c^2}\right)^2\left(\frac{Z}{137}\right)^3\frac{1}{f(Z_d, E_0)}.$$

Note that $E_\nu \approx E_0 + m_e c^2$, and $e^2/4\pi\varepsilon_0\hbar c \approx \frac{1}{137}$. For low values of Z this ratio is usually small, but at high Z the Z^3 factor, and the increasing Coulomb barrier for positron emission which reduces $f(Z_d, E_0)$, make electron capture the dominant process.

12.5 The neutrino mass

In standard unified theories the neutrino mass is assumed to be zero. It is clearly important to test this assumption experimentally. A finite neutrino mass of even a few eV/c^2 would have significant consequences in, for example, cosmology. The signature of a finite neutrino mass would in

β-decay appear in the shape of the electron energy spectrum near maximum energy. From expression (12.5), this shape depends sensitively on whether $m_\nu = 0$ or $m_\nu \neq 0$. The difference is clearer in a *Kurie plot* of

$$\left[\frac{\mathrm{d}R/\mathrm{d}E_\mathrm{e}}{F(Z_\mathrm{d}, E_\mathrm{e})E_\mathrm{e}(E_\mathrm{e}^2 - m_\mathrm{e}^2c^4)^{\frac{1}{2}}}\right]^{\frac{1}{2}}$$

against electron energy since from equation (12.5)

$$\left[\frac{\mathrm{d}R/\mathrm{d}E_\mathrm{e}}{F(Z_\mathrm{d}, E_\mathrm{e})E_\mathrm{e}(E_\mathrm{e}^2 - m_\mathrm{e}^2c^4)^{\frac{1}{2}}}\right]^{\frac{1}{2}} = (\mathrm{constant})(E_0 - E_\mathrm{e})^{\frac{1}{2}}[(E_0 - E_\mathrm{e})^2 - m_\nu^2c^4]^{\frac{1}{4}}.$$

If $m_\nu = 0$, this plot gives a straight line $(E_0 - E_\mathrm{e})$ passing through E_0; if $m_\nu \neq 0$ the line is curved and the tangent at maximum energy is vertical.

A much-studied decay in this context is that of tritium

$$^3_1\mathrm{H} \rightarrow {}^3_2\mathrm{He} + \mathrm{e}^- + \bar{\nu} + 18.6\ \mathrm{keV}.$$

The low electron kinetic energies in this decay are experimentally advantageous. Figure 12.4 shows experimental data and there is remarkable overall agreement between the data and the fitted theoretical spectrum. A Kurie plot of data near $E_\mathrm{e} = E_0$ is also shown. The difficulty of the experiment is evident: the conclusion is that $m_\nu < 60\ \mathrm{eV}/c^2$. More recent (1980) data from the Soviet Union with a bigger sample of decays indicates that the neutrino mass is finite and perhaps as large as 34 eV/c^2, but there remain uncertainties in the interpretation of the results.

12.6 The Fermi and Gamow–Teller interactions

In the simple Fermi theory of β-decay, the interaction matrix element was written as a 'contact' interaction. For our example of $^{17}\mathrm{F}$ decay,

$$H_{\mathrm{f}0} = \int \Psi_\mathrm{f}^* H \Psi_\mathrm{i}\ \mathrm{d(coordinates)}$$

$$= G_\mathrm{F} \int \psi_\mathrm{n}^*(\mathbf{r})\psi_\mathrm{e}^*(\mathbf{r})\psi_\nu^*(\mathbf{r})\psi_\mathrm{p}(\mathbf{r})\ \mathrm{d}^3\mathbf{r},$$

which we might represent diagrammatically as in Fig. 12.5. Reference to spin has been suppressed, though we know that the particles involved are all fermions with intrinsic spin quantum number $s = \frac{1}{2}$.

In the full theory of β-decay, the interaction is mediated by the charged W bosons, so that the process above is represented by Fig. 12.6. At a more fundamental level, the interaction is with a quark rather than a nucleon, as in Fig. 3.3, but phenomenologically the principal missing feature of the simple Fermi theory is the description of spin effects. We now include spin but we shall restrict the discussion to allowed transitions. The nucleon states in our example can be described non-relativistically by $\psi_p(\mathbf{r}_p, m_{sp})$, $\psi_n(\mathbf{r}_n, m_{sn})$ where we now explicitly include the intrinsic spin (m_s can take on the values $\pm\frac{1}{2}$). However, both the positron and neutrino move at relativistic speeds and for these the relativistic wave-functions must be used. Except in terms of the Dirac wave-functions, there is no simple form for the lepton part of the matrix element.

The contribution to the interaction from the Coulomb-like part of the

12.4 The electron energy spectrum from the decay of tritium. The experimental points give the number of electrons $N(E_e)$ observed in small energy 'bins' from a very large number of decays. (Taken from Lewis, V. E. (1970), *Nuc. Phys.* **A151**, 120.) The spectrum is well fitted using equations (12.5) and (12.6). Also shown for comparison is the curve without the Coulomb correction. The inset shows a Kurie plot of the spectrum near the electron end point. (For this data see Bergkuist, K. E. (1972), *Nuc. Phys.* **B39**, (317), and for more recent data Lubimov, V. A. *et al.* (1980), *Phys. Lett.* **94B**, 266.) The theoretical curves in the inset include the effect of the finite size of the energy 'bins'.

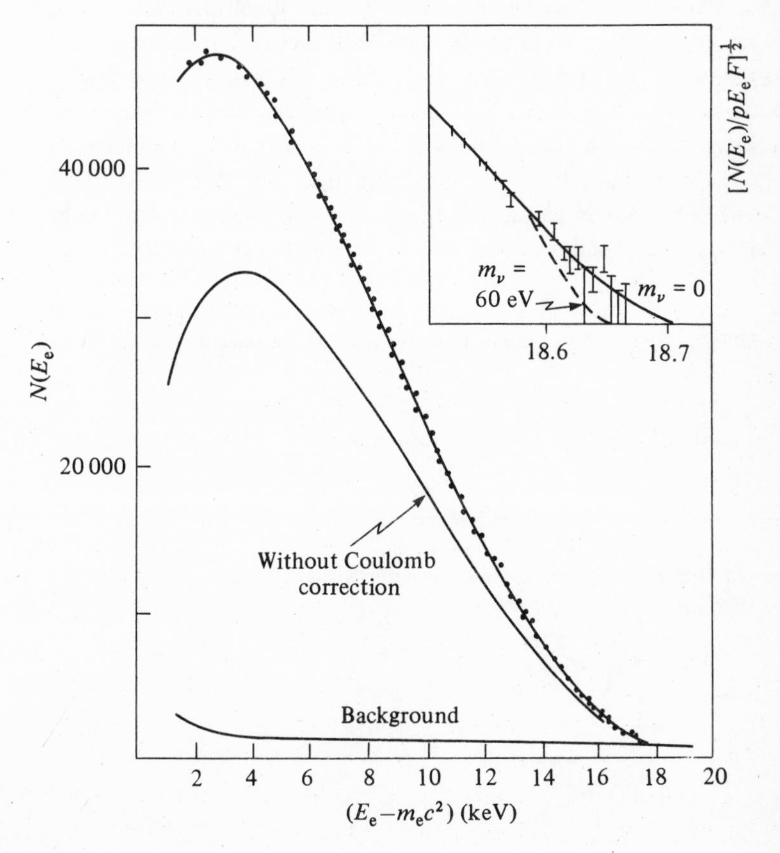

W-field is most like the simple Fermi theory discussed in the previous sections and it is called the Fermi interaction. This part does not change the nucleon spins, and for allowed transitions the positron and neutrino angular momenta must combine to give a total lepton angular momentum of zero. There is a coupling constant g_V associated with the nucleon–W interaction in Fig. 12.6 and a coupling constant g_L associated with the lepton–W interaction. Both g_V and g_L have dimensions of electric charge. The heavy mass of the W boson makes the interaction short-ranged and introduces a factor $(1/\varepsilon_0)(\hbar/M_W c)^2$, as after equation (2.13).

Thus the contribution of the Fermi interaction to the interaction matrix element is

$$H_{f0}^{F} = g_V g_L \frac{1}{\varepsilon_0}\left(\frac{\hbar}{M_W c}\right)^2 \int \psi_n^*(\mathbf{r}, m_s)\psi_p(\mathbf{r}, m_s)\, d^3\mathbf{r}\, dm_s \quad \text{(lepton part).}$$

(The dm_s stands for a sum over the spin coordinates.)

The subtlety of the weak interaction is contained in the bracketed lepton part. This involves the neutrino and positron wave-functions evaluated at the nucleon coordinate **r**, as in the simple Fermi theory, but also describes the alignment of the neutrino and positron spins and the angular correlation between their directions. (The neutrino direction can be inferred by measuring the small nuclear recoil.) However, an experiment which only measures the electron energy spectrum and does not distinguish these correlations corresponds to an averaging over directions, and then the spectrum is given exactly as in the simple theory. If only the Fermi

12.5 β-decay of a proton in a nucleus as a 'contact' interaction.

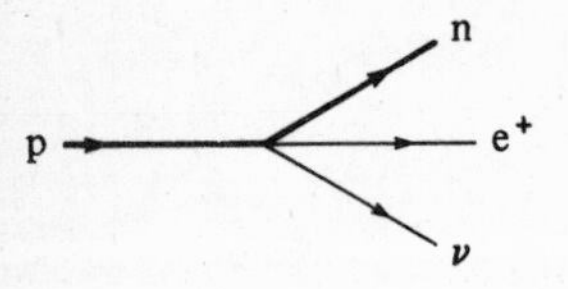

12.6 β-decay of a proton in a nucleus mediated by the exchange of a virtual W boson.

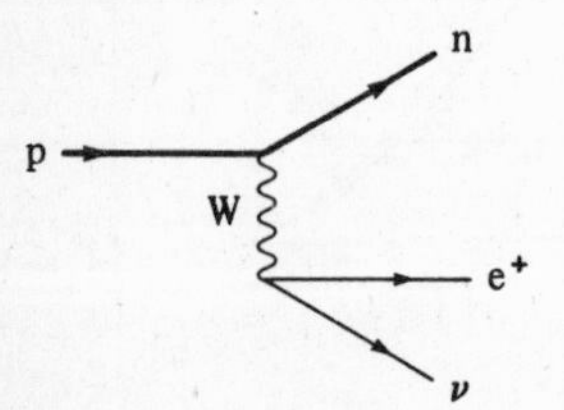

interaction contributes to the decay we can make the identifications

$$G_{\mathrm{F}} = g_{\mathrm{V}} g_{\mathrm{L}} \frac{1}{\varepsilon_0} \left(\frac{\hbar}{M_{\mathrm{W}} c} \right)^2,$$

$$M_{\mathrm{F}} = \int \psi_{\mathrm{n}}^*(\mathbf{r}, m_{\mathrm{s}}) \psi_{\mathrm{p}}(\mathbf{r}, m_{\mathrm{s}})\, \mathrm{d}^3\mathbf{r}\, \mathrm{d}m_{\mathrm{s}}.$$

But this is not the whole story, even for allowed transitions. The magnetic-like part of the W-field leads to a term in the transition matrix element, known as the Gamow–Teller interaction, in which the total lepton angular momentum **J** has quantum number $j = 1$, and the nuclear part of the interaction (again treated non-relativistically) contains the Pauli operator $\boldsymbol{\sigma}$ (see Appendix C). There is a term $\boldsymbol{\sigma} \cdot \mathbf{J}$ in the interaction Hamiltonian. An 'axial coupling constant' g_{A} takes the place of g_{V}. The Gamow–Teller matrix element for our ^{17}F example is

$$H_{\mathrm{f0}}^{\mathrm{GT}} = g_{\mathrm{A}} g_{\mathrm{L}} \frac{1}{\varepsilon_0} \left(\frac{\hbar}{M_{\mathrm{W}} c} \right)^2 \int \psi_{\mathrm{n}}^*(\mathbf{r}, m_{\mathrm{s}}) \boldsymbol{\sigma} \psi_{\mathrm{p}}(\mathbf{r}, m_{\mathrm{s}})\, \mathrm{d}^3\mathbf{r}\, \mathrm{d}m_{\mathrm{s}} \cdot (\text{lepton part})$$

$$= \frac{g_{\mathrm{A}}}{g_{\mathrm{V}}} G_{\mathrm{F}} \int \psi_{\mathrm{n}}^*(\mathbf{r}, m_{\mathrm{s}}) \boldsymbol{\sigma} \psi_{\mathrm{p}}(\mathbf{r}, m_{\mathrm{s}})\, \mathrm{d}^3\mathbf{r}\, \mathrm{d}m_{\mathrm{s}} \cdot (\text{lepton part}),$$

where (lepton part) is now a vector and $\cdot$ denotes a scalar product.

If we define

$$\mathbf{M}_{\mathrm{GT}} = \frac{g_{\mathrm{A}}}{g_{\mathrm{V}}} \int \psi_{\mathrm{n}}^*(\mathbf{r}, m_{\mathrm{s}}) \boldsymbol{\sigma} \psi_{\mathrm{p}}(\mathbf{r}, m_{\mathrm{s}})\, \mathrm{d}^3\mathbf{r}\, \mathrm{d}m_{\mathrm{s}} = (M_{\mathrm{GT}}^x, M_{\mathrm{GT}}^y, M_{\mathrm{GT}}^z),$$

and sum over all allowed decays to the $j = 0$ and the three $j = 1$ states, the total decay rate to electrons with energies in the range E, $E + \mathrm{d}E$ is given by

$$\mathrm{d}R(E_{\mathrm{e}}) = \frac{G_{\mathrm{F}}^2}{2\pi^3 \hbar^7 c^6} [|M_{\mathrm{F}}|^2 + |M_{\mathrm{GT}}^x| + |M_{\mathrm{GT}}^y|^2 + |M_{\mathrm{GT}}^z|^2] S_{\mathrm{c}}(E_{\mathrm{e}})\, \mathrm{d}E_{\mathrm{e}}, \tag{12.8}$$

and the mean life is given by

$$\frac{1}{\tau} = \frac{G_{\mathrm{F}}^2 m_{\mathrm{e}}^5 c^4}{2\pi^3 \hbar^7} [|M_{\mathrm{F}}|^2 + |M_{\mathrm{GT}}^x|^2 + |M_{\mathrm{GT}}^y|^2 + |M_{\mathrm{GT}}^z|^2] f(Z_{\mathrm{d}}, E_0).$$

An allowed decay may be pure Fermi, pure Gamow–Teller, or a mixture of both, depending on the details of the nuclear matrix elements. Note that the electron energy spectrum is independent of these details. In general, of course, the initial and final nuclear states which enter into M_{F} and $\mathbf{M}_{\mathrm{GT}}$ are more complicated than those of our ^{17}F example. M_{F} and $\mathbf{M}_{\mathrm{GT}}$ always vanish if the initial and final nuclear states are of opposite parity, since $\boldsymbol{\sigma}$ is an axial vector. Thus there can be no parity change in the nuclear states in an allowed transition.

For a Fermi transition, the change Δj in nuclear spin must be zero. For a Gamow–Teller transition, $\Delta j = 0$ or 1, by the rules for addition of angular momentum, except that $0 \to 0$ transitions are forbidden since the matrix element of $\boldsymbol{\sigma}$ vanishes between two spherically symmetrical states.

12.7 The determination of the coupling constants

In the decay

$$^{14}_{8}\mathrm{O} \to {}^{14}_{7}\mathrm{N}^* + \mathrm{e}^+ + \nu$$

the transition occurs, with 99.7% probability, to the first excited state of the daughter nucleus, which has spin and parity 0^+. The even–even nucleus $^{14}_{8}\mathrm{O}$ also has spin and parity 0^+, so that from the selection rules above the transition is allowed, and pure Fermi. Also, in the nuclear shell model, the nuclei differ only in that $^{14}\mathrm{O}$ has two protons in $1\mathrm{p}_{\frac{1}{2}}$ states outside a $^{12}_{6}\mathrm{C}$ core, and $^{14}\mathrm{N}^*$ has one proton and one neutron. Thus, because of the charge independence of the strong nuclear force and the smallness of the Coulomb effects in these light nuclei, the wave-functions of the initial and final nuclear states are very similar, and $|M_{\mathrm{F}}|^2 \approx 2$ (since either of the two protons in $1\mathrm{p}_{\frac{1}{2}}$ states can decay). The energy $E_0 = 2.32$ MeV, $Z_{\mathrm{d}} = 7$ and $f(7, 2.32) = 42.8$. The measured mean life is 102 s. Thus from the formula (12.7) for the mean life we can calculate

$$G_{\mathrm{F}}/(\hbar c)^3 = 1.16 \times 10^{-5}\ \mathrm{GeV}^{-2}.$$

This mean life measurement (1954) gave one of the first clean estimates of the Fermi coupling constant, and the value found is close to the best value, $G_{\mathrm{F}}/(\hbar c)^3 = 1.136 \times 10^{-5}\ \mathrm{GeV}^{-2}$, known today.

The ratio $g_{\mathrm{A}}/g_{\mathrm{V}}$ is most directly determined from the lifetime of a free neutron, since there are then no uncertainties in the computation of nuclear wave-functions. Indeed, if we neglect recoil, the spatial parts of the initial neutron and final proton wave-functions are the same. Suppose the spin state of the neutron is $|+\frac{1}{2}\rangle$. If the proton spin state is $|+\frac{1}{2}\rangle$ then, using the properties of the $\boldsymbol{\sigma}$ matrices (Appendix C) $M_{\mathrm{F}} = 1$ and $\mathbf{M}_{\mathrm{GT}} = (g_{\mathrm{A}}/g_{\mathrm{V}})(0, 0, 1)$. If the proton spin state is $|-\frac{1}{2}\rangle$, $M_{\mathrm{F}} = 0$ and $\mathbf{M}_{\mathrm{GT}} = (g_{\mathrm{A}}/g_{\mathrm{V}})(1, i, 0)$. The neutron–proton mass difference gives $E_0 = 1.29$ MeV and $f(1, 1.29) = 1.63$. The total decay rate to all possible spin states is therefore, from equations (12.6) and (12.8) given by

$$\frac{1}{\tau} = \frac{1.63[1 + 3g_{\mathrm{A}}^2/g_{\mathrm{V}}^2]G_{\mathrm{F}}^2 m^5 c^4}{2\pi^3 \hbar^7}$$

and the measured decay rate yields

$$g_{\mathrm{A}}/g_{\mathrm{V}} = 1.26.$$

For calculations of β-decay properties, the values of G_F and g_A/g_V are sufficient, but the coupling constants g_L, g_V are related to the fundamental electric charge e by two constants of particle physics, the Weinberg angle θ_W, and the Cabibbo angle θ_C:

$$g_L = \frac{e}{2 \sin \theta_W} = (1.06 \pm 0.04)e,$$

$$g_V = \frac{e \cos \theta_C}{2\sqrt{2} \sin \theta_W} = (0.73 \pm 0.03)e.$$

The Weinberg angle appears in the standard electro-weak theory. To a good approximation the masses of the W and Z bosons are related to θ_W by:

$$\left(\frac{M_W}{M_Z}\right)^2 = \cos^2 \theta_W.$$

The Cabibbo angle θ_C describes, at a deeper level of the theory, the ratio of coupling constants of different quark types. Empirically, θ_C is nearly zero and $\cos \theta_C = 0.9737 \pm 0.0045$.

At the quark level, the coupling constants g_V, g_A are thought to be the same, but g_A is modified in a neutron because of quark interactions. In principle there could be corrections to the free neutron value for β-decays in nuclei.

12.8 Electron polarisation

The lepton part of the interaction matrix element leads to angular correlations between the various spins and momenta of the four particles involved in a β-decay. These correlations can be detected in suitable experiments, as for example the spin-polarised ^{60}Co experiment discussed in § 12.1; the observed angular distribution of electrons in this experiment is in accord with the theory.

The non-parity conserving nature of the weak interaction is most clearly seen in the lepton states. All neutrinos are 'left-handed' and all anti-neutrinos 'right-handed'. The theory also predicts that in β-decay left-handed electrons are produced more copiously than right-handed electrons, whereas positrons produced in β-decay are predominantly right-handed. More precisely, the probability of an electron emitted with velocity v being in a left-handed state (with intrinsic spin **s** anti-parallel to momentum **p**) is

$$P_L = \frac{1}{2}\left(1 + \frac{v}{c}\right),$$

and the probability of its being emitted in a right-handed state (with **s** parallel to **p**) is

$$P_{\mathrm{R}}=\frac{1}{2}\left(1-\frac{v}{c}\right).$$

Hence

$$P=\frac{P_{\mathrm{R}}-P_{\mathrm{L}}}{P_{\mathrm{R}}+P_{\mathrm{L}}}=-\frac{v}{c}$$

(and for positrons $P=+v/c$).

Figure 12.7 shows experimental measurements of P, the 'longitudinal polarisation', plotted against v/c for a variety of β-decays. The complete polarisation of the neutrino and anti-neutrino can be regarded as a generalisation of this result, since $v=c$ for massless particles.

12.9 Theory of γ-decay

In γ-decay, a nucleus in an excited state falls to a lower state with the emission of a photon (§7.3). The electromagnetic interaction which governs this process is very well understood theoretically, but a full discussion requires the quantised equations of the electromagnetic field,

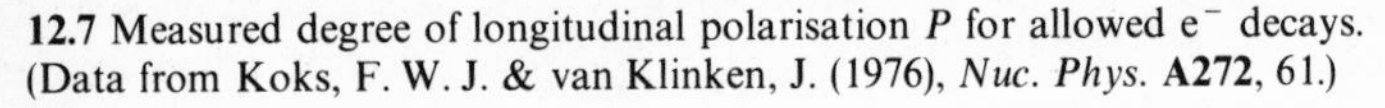

12.7 Measured degree of longitudinal polarisation P for allowed e^- decays. (Data from Koks, F. W. J. & van Klinken, J. (1976), *Nuc. Phys.* **A272**, 61.)

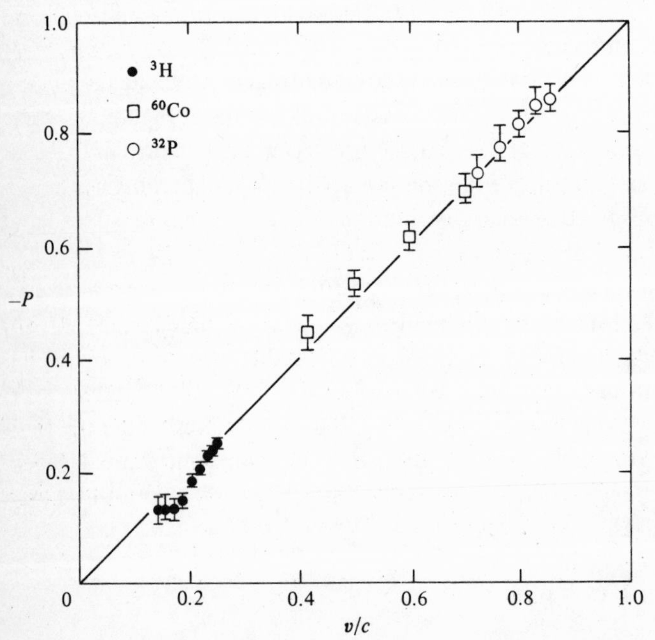

rather than the classical Maxwell equations, and is beyond the scope of this book. However, we can understand the main features of γ-decay, and in particular the great range of γ-decay lifetimes described in § 7.3, using semi-classical arguments to write down an approximate expression for the interaction energy between a nucleus and a photon.

We again enclose our system in a large volume V. Consider the plane electromagnetic wave

$$\mathbf{E}=\mathbf{E}_0 \cos(\mathbf{k}\cdot\mathbf{r}-\omega t), \quad \mathbf{B}=\mathbf{B}_0 \cos(\mathbf{k}\cdot\mathbf{r}-\omega t).$$

The standard Maxwell theory tells us that in such a wave $|\mathbf{B}|=|\mathbf{E}|/c$, and the energy is divided equally between the electric and magnetic fields and is given in total by

$$\varepsilon_0 \int \mathbf{E}(\mathbf{r})\cdot\mathbf{E}(\mathbf{r})\, \mathrm{d}^3\mathbf{r}=\tfrac{1}{2}\varepsilon_0 \mathbf{E}_0^2 V,$$

since the cosine squared averages to $\frac{1}{2}$ over the volume V. If we identify this wave with a single photon of wave-vector $\mathbf{k}$ and energy $\hbar\omega$ we must therefore set

$$\tfrac{1}{2}\varepsilon_0 \mathbf{E}_0^2 V=\hbar\omega, \quad \text{or} \quad |\mathbf{E}_0|=(2\hbar\omega/\varepsilon_0 V)^{\frac{1}{2}}. \tag{12.9}$$

In a typical γ-decay, $\hbar\omega$ is at most a few MeV, so that $|\mathbf{k}|=\hbar\omega/\hbar c\sim$ (1 MeV)/(197 MeV fm) $\sim 10^{-2}$ fm^{-1}. Hence to a good approximation we can neglect the change in $(\mathbf{k}\cdot\mathbf{r})$ over the dimensions of the nucleus ($\sim$fm), which we can take to be centred at $\mathbf{r}=0$. The electric field over the nucleus is then

$$\mathbf{E}=\mathbf{E}_0 \cos\omega t=\tfrac{1}{2}\mathbf{E}_0(e^{i\omega t}+e^{-i\omega t}),$$

and the potential energy of the nucleus in such a field is given classically by

$$-e\sum_{\text{protons}} \mathbf{E}\cdot\mathbf{r}_\mathrm{p}=-\frac{e}{2}\sum_{\text{protons}} \mathbf{E}_0\cdot\mathbf{r}_\mathrm{p}(e^{i\omega t}+e^{-i\omega t}).$$

In a γ-decay, we start with a state in which there is no photon present, and end with a state in which there is one photon present and the nucleus is in a lower energy state. As in our discussion of β-decay, we shall neglect the small nuclear recoil energy. It is clear from the derivation of the result (D.6) in Appendix D that only the term with $e^{i\omega t}$ in the interaction can contribute to this transition, so that the matrix element to be employed in the formula for the decay rate is

$$-\frac{e}{2}\int \Psi_\mathrm{f}^*\Big(\sum_{\text{protons}} \mathbf{E}_0\cdot\mathbf{r}_\mathrm{p}\Big)\Psi_0\, \mathrm{d(coordinates)}=-\frac{e}{2}\mathbf{E}_0\cdot\mathbf{R}_{\mathrm{f}0},$$

where Ψ_0, Ψ_f are the initial and final nuclear states and

$$\mathbf{R}_{f0} = \int \Psi_f^* \left(\sum_{\text{protons}} \mathbf{r}_p \right) \Psi_0 \, \mathrm{d(coordinates)}. \qquad (12.10)$$

If $\mathbf{R}_{f0}$ is non-vanishing the transition is said to be *electric dipole* (E1).

Let us assume $\mathbf{R}_{f0}$ is non-vanishing and real. The treatment is easily extended to the case when $\mathbf{R}_{f0}$ is a complex vector $\mathbf{R}_1 + i\mathbf{R}_2$ since

$$|\mathbf{E}_0 \cdot (\mathbf{R}_1 + i\mathbf{R}_2)|^2 = |\mathbf{E}_0 \cdot \mathbf{R}_1|^2 + |\mathbf{E}_0 \cdot \mathbf{R}_2|^2.$$

For a given direction of photon emission, there are two independent photon states with polarisations which we can take as in the plane defined by $\mathbf{R}_{f0}$ and $\mathbf{k}$, and perpendicular to this plane (Fig. 12.8). For the latter, $\mathbf{E}_0 \cdot \mathbf{R}_{f0} = 0$, so that the transition probability to this state vanishes.

If θ is the angle between the direction $\mathbf{k}$ and $\mathbf{R}_{f0}$, for the state with polarisation in the plane we have

$$\mathbf{E}_0 \cdot \mathbf{R}_{f0} = |\mathbf{E}_0| \, |\mathbf{R}_{f0}| \sin \theta$$

since $\mathbf{E}_0$ is perpendicular to $\mathbf{k}$. The density of states at energy $E_\gamma = \hbar\omega$, for photons emitted in a solid angle $\mathrm{d}\Omega = \sin \theta \, \mathrm{d}\theta \, \mathrm{d}\phi$, is

$$\frac{V}{(2\pi)^3} k^2 \frac{\mathrm{d}k}{\mathrm{d}E_\gamma} \mathrm{d}\Omega = \frac{V}{(2\pi)^3} \frac{\omega^2}{\hbar c^3} \sin \theta \, \mathrm{d}\theta \, \mathrm{d}\phi,$$

since $E_\gamma = \hbar\omega = \hbar ck$. Hence the 'Fermi golden rule' formula gives the transition rate

$$\frac{2\pi}{\hbar} \frac{e^2}{4} |\mathbf{E}_0|^2 |\mathbf{R}_{f0}|^2 \frac{V}{(2\pi)^3} \frac{\omega^2}{\hbar c^3} \sin^2 \theta \, \mathrm{d}\Omega. \qquad (12.11)$$

There is a characteristic $\sin^2 \theta$ angular distribution of the emitted photons. Such angular distributions can be observed experimentally, if for example the nuclei in a sample are oriented in the same direction by a strong magnetic field.

The mean life is obtained by integrating the expression (12.11) over all

12.8 Direction of emission **k** and polarisation vector $\mathbf{E}_0$ for an allowed electric dipole transition.

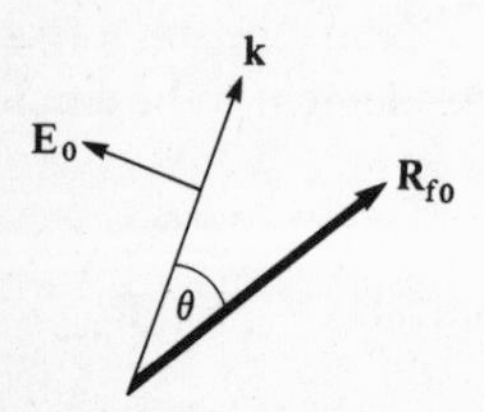

directions in space. Using the value of $|\mathbf{E}_0|$ given by (12.9) we obtain

$$\frac{1}{\tau_{E1}}=\frac{4}{3}\left(\frac{e^2}{4\pi\varepsilon_0}\right)\frac{\omega^3}{\hbar c^3}|\mathbf{R}_{f0}|^2$$
$$=\left(\frac{E_\gamma}{1\,\mathrm{MeV}}\right)^3\left(\frac{|\mathbf{R}_{f0}|}{1\,\mathrm{fm}}\right)^2 \quad 0.38\times10^{15}\,\mathrm{s}^{-1}, \tag{12.12}$$

where the last form indicates the order of magnitude to be expected for the mean lives.

From equation (12.10), we see that Ψ_0 and Ψ_f must be of opposite parity for an electric dipole transition to take place, since if they have the same parity the integral vanishes. It can also be shown from the angular part of the integration, using the properties of spherical harmonics, that the change Δj in the nuclear spin quantum number for an electric dipole transition must be $\Delta j=0$ or $\Delta j=\pm1$, except that $0\to0$ transitions are forbidden. An estimate of the magnitude of $|\mathbf{R}_{f0}|$ requires a knowledge of the nuclear wave functions. Even in the simple shell model such calculations are not easy.

The nucleus also couples to the magnetic field of the photon, and at a similar level of approximation the interaction with the magnetic field $\mathbf{B}$ is given classically by $-\boldsymbol{\mu}\cdot\mathbf{B}_0\cos\omega t$, where $\boldsymbol{\mu}$ is the total magnetic moment of the nucleus. The magnetic moment operator is given in the simple shell model by equation (5.24),

$$\boldsymbol{\mu}=\sum_{\text{nucleons}}\mu_N[g_L\mathbf{L}+g_s\mathbf{s}]/\hbar,$$

where $\mu_N=e\hbar/2m_p$ is the nuclear magneton. The transition rate induced by this interaction will be of the same form as (2.11), with $e\mathbf{E}_0\cdot\mathbf{R}_{f0}$ replaced by $\mathbf{B}_0\cdot\mathbf{M}_{f0}$ where

$$\mathbf{M}_{f0}=\int\Psi_f^*\boldsymbol{\mu}\Psi_0\,\mathrm{d(coordinates)}. \tag{12.13}$$

If $\mathbf{M}_{f0}$ is non-vanishing, the transition is said to be *magnetic dipole* (M1). Since $|\mathbf{B}_0|=|\mathbf{E}_0|/c$, the mean life is given by

$$\frac{1}{\tau_{M1}}=\frac{4}{3}\left(\frac{1}{4\pi\varepsilon_0}\right)\frac{\omega^3}{\hbar^2c^5}|\mathbf{M}_{f0}|^2. \tag{12.14}$$

From (12.13), $\mathbf{M}_{f0}$ is non-vanishing only if Ψ_0 and Ψ_f have the same parity, since $\boldsymbol{\mu}$ is a pseudo-vector (§2.6). Electric and magnetic dipole transitions are therefore mutually exclusive. The angular momentum selection rules, $\Delta j=0$, ±1 ($0\to0$ forbidden) are the same as for electric dipole transitions.

The ratio of mean lives for magnetic and electric dipole transitions at the same energy is

$$\frac{\tau_{\mathrm{M1}}}{\tau_{\mathrm{E1}}}=\frac{e^2c^2|\mathbf{R}|^2}{|\mathbf{M}|^2}.$$

If we take $|\mathbf{R}|\sim$ nuclear radius $\sim A^{\frac{1}{3}}$ fm, and $M\sim e\hbar/m_{\mathrm{p}}$, we obtain

$$\frac{\tau_{\mathrm{M1}}}{\tau_{\mathrm{E1}}}\sim\frac{(m_{\mathrm{p}}c^2)^2(A^{\frac{2}{3}}\,\mathrm{fm}^2)}{(\hbar c)^2}\sim 20A^{\frac{2}{3}}.$$

Thus the mean lives for magnetic dipole transitions are generally longer than those of electric dipole transitions at the same energy by a considerable factor, though this estimate is of course very crude.

If both $\mathbf{R}_{\mathrm{f0}}$ and $\mathbf{M}_{\mathrm{f0}}$ vanish, as is not uncommon, then we can no longer neglect the variation in the photon field over the dimensions of the nucleus. The expansion of $\cos(\mathbf{k}\cdot\mathbf{r}-\omega t)$ in powers of $(\mathbf{k}\cdot\mathbf{r})$ gives matrix elements for higher-order electric and magnetic transitions. Each power of $(\mathbf{k}\cdot\mathbf{r})$ introduces an additional factor of -1 in the parity selection rule, and an additional unit of orbital angular momentum in the Δj selection rule so that, for example, for electric quadrupole transitions there is no change in parity and $\Delta j=\pm 2, \pm 1, 0$, except that $0\rightarrow 0$ and $\frac{1}{2}\rightarrow\frac{1}{2}$ transitions are forbidden. To each type of transition there corresponds a characteristic angular dependence and polarisation of emitted γ-rays.

Each power of $(\mathbf{k}\cdot\mathbf{r})$ reduces the order of magnitude of the matrix element by a factor $\sim(kR)$, where R is the nuclear radius, and hence increases the lifetime by a factor of $(kR)^{-2}$. For a 1 MeV photon and $A\sim 50$, $(kR)^{-2}=0.24\times 10^4$. The curves of Fig. 7.6 have been drawn using only a more sophisticated version of this argument, but they are nevertheless a useful guide to the interpretation of experimental lifetimes.

12.10 Internal conversion

A nucleus in an excited state can also decay electromagnetically by 'internal conversion'. In this process, an atomic electron in a state $\phi_0(\mathbf{r}_{\mathrm{e}})$ takes up the energy released in the decay and is excited to a state $\phi_{\mathrm{f}}(\mathbf{r}_{\mathrm{e}})$ which must be initially empty. If the energy release is greater than the binding energy of the electron, as is usually the case, the electron is ejected from the atom and the state $\phi_{\mathrm{f}}(\mathbf{r}_{\mathrm{e}})$ may be approximated by a plane wave.

Thus the initial state is of the form

$$\Psi_0=\psi_0^{\mathrm{nuc}}\phi_0(\mathbf{r}_{\mathrm{e}})$$

and the final state of the form

$$\Psi_{\mathrm{f}}=\psi_{\mathrm{f}}^{\mathrm{nuc}}\phi_{\mathrm{f}}(\mathbf{r}_{\mathrm{e}}).$$

The main contribution to the interaction energy between the electron and the nucleus is the Coulomb energy

$$\sum_{\text{protons}} \frac{-e^2}{4\pi\varepsilon_0|\mathbf{r}_p-\mathbf{r}_e|}$$

and the corresponding matrix element for the transition is

$$H_{f0}=\int \psi_f^{\text{nuc}*}\phi_f^* \sum_{\text{protons}} \frac{-e^2}{4\pi\varepsilon_0|\mathbf{r}_p-\mathbf{r}_e|} \times \psi_0^{\text{nuc}}\phi_0 \, d^3\mathbf{r}_e \, d(\text{nuclear coordinates}).$$

We shall not pursue the evaluation of this matrix element, but note that it can be non-vanishing for $0 \to 0$ transitions between nuclear states of the same parity.

The process of internal conversion always competes with γ-decay with similar nuclear matrix elements appearing. As in the case of K-capture in β-decay, there is a factor Z^3 in the transition rate arising from the normalisation of the initial state electron wave-function. Thus internal conversion becomes increasingly significant in the electromagnetic decays of the heavier elements. The *internal conversion coefficient* is defined as the ratio of the rate of internal conversion to the rate of γ-emission, for a given type of electromagnetic transition. Extensive tables of these coefficients can be found in the literature.

Problems

12.1 Consider allowed β-decays which have a large energy release E_0 (e.g. the decay of ^{8}B, § 10.4). In such decays, the effects of Coulomb corrections and finite lepton masses are small. Show that, neglecting these effects,

(*a*) the mean life depends on E_0 as E_0^{-5},

(*b*) the mean electron energy is $E_0/2$.

To examine the effect of a finite neutrino mass on the energy spectrum, only decays with energy in a small range $\Delta E_e \sim m_\nu c^2$ at the end-point $E_e \approx E_0$ are significant. Show that the proportion of such decays is very small, of order $10(\Delta E_e/E_0)^3$.

12.2 In the K-capture

$^7_4$Be (atom) $\to$ $^7_3$Li (atom) $+ \nu$,

with the beryllium source at rest, the recoil energy of the lithium atoms (mass 6536 MeV/c^2) was measured to be (55.9 ± 1.0) eV (Davis, R. (1952), *Phys. Rev.* **86**, 976). The mass difference between the two atoms is 0.862 MeV/c^2. Show that this experiment implied the neutrino mass to be less than 160 keV/c^2.

12.3 The product of a β-decay half life $T_{\frac{1}{2}}$ (§2.4) and the number $f(Z_d, E_0)$ is the '$fT_{\frac{1}{2}}$' value. From equations (12.7) and (12.8) the $fT_{\frac{1}{2}}$ value gives a direct empirical determination of the nuclear matrix element $|M_F|^2+|\mathbf{M}_{GT}|^2$. Calculate the $fT_{\frac{1}{2}}$ value for the decay ${}^{31}_{16}S \to {}^{31}_{15}P + e^+ + \nu$, for which $T_{\frac{1}{2}}=2.60$ s, $E_0=4.94$ MeV and $f(15, 4.94)=1830$. In the simple shell model, this decay involves a $2s_{\frac{1}{2}}$ proton changing to a $2s_{\frac{1}{2}}$ neutron. Compare this $fT_{\frac{1}{2}}$ value with that of a free neutron. Why do the two values differ?

12.4 The cross-section for the reaction

$$\bar{\nu}+p \to n+e^+$$

is given in perturbation theory by

$$\sigma=\frac{2\pi}{\hbar}\frac{1}{(\text{neutrino flux})}|H_{f0}|^2 n(E_e),$$

where $n(E_e)$ is the relativistic density of states (equation (12.4)). Show that

$$\sigma \approx \frac{G_F^2}{\pi(\hbar c)^4}\left[1+\frac{3g_A^2}{g_V^2}\right](cp_e)E_e.$$

Calculate this cross-section for a 2 MeV anti-neutrino.

12.5 If an electric dipole (E1) decay mean life is known, then equation (12.12) can be used to calculate the corresponding dipole matrix element $|\mathbf{R}_{f0}|$. The first excited state of ^{11}Be decays to the ground state through an E1 transition. The mean life is 1.79×10^{-13} s and the photon energy is 0.32 MeV. Calculate $|R_{f0}|$.

An example of an electric dipole transition in atomic physics is the decay of the 2p excited state of the hydrogen atom, for which the mean life is 1.6×10^{-9} s and the photon energy is 10.2 eV. Calculate $|\mathbf{R}_{f0}|$ and compare it with the nuclear matrix element.

12.6 The nucleus ${}^{108}_{47}$Ag, which has spin and parity 1^+, is β-unstable with a mean life of 3.4 minutes. It has an excited state at 109 keV excitation energy, spin and parity 6^+, which is an isomeric state with a mean life of 180 years. Explain how an excited state of a nucleus can be more stable than the ground state.

13 The passage of energetic particles through matter

As a coda to this book, we consider the passage of energetic particles through matter. Nuclear reactions usually result in the production of such particles: α-particles, electrons, photons, nucleons, fission fragments, or whatever. In passing through matter, an energetic particle loses its energy, ultimately largely into ionisation. The instruments of nuclear physics are designed to detect and measure this deposited energy, and so it is upon these processes that our knowledge of nuclear physics rests.

The subject is also basic to an understanding of the biological effects of energetic particles, since a living cell can be damaged by the ionisation. This can be of positive benefit, as in the destruction of malignant tissue in cancer treatment, or a danger from which, for example, workers in the nuclear power industry must be shielded. Shielding calculations also depend on the physical principles set out in this chapter.

We limit the discussion to particles with kinetic energies up to around 10 MeV, in line with the nuclear physics described in Chapters 4–12. It is intended to give the reader a qualitative comprehension, rather than a compendium of the most accurate formulae and data available for quantitative work.

13.1 Charged particles

We consider first the passage of charged particles, such as protons and α-particles, through gases. For charged particles of energy <10 MeV, the dominant mechanism for energy loss is the excitation or ionisation of the atoms (or molecules) of the gas: electrons being excited to higher bound energy levels in the atom, or detached completely. The essential physics of the process may be understood using classical mechanics.

We consider a 'fast' particle, charge ze, velocity v, energy E, passing a particle of charge $z'e$, mass m_R, initially at rest. We suppose that the fast particle deviates a negligible amount from its initial straight-line path along the x-axis (Fig. 13.1), and the rest particle at the point $(0, b, 0)$ moves only a negligible distance during the encounter. The distance b is called the *impact parameter*.

The equation of motion of the fast particle is

$$\frac{\mathrm{d}\mathbf{p}}{\mathrm{d}t}=ze\mathbf{E},$$

where $\mathbf{p}$ is its momentum and $\mathbf{E}$ is the electric field due to the rest particle. The magnetic field due to the 'rest' particle will be negligible. This equation remains valid for relativistic momenta.

The field $\mathbf{E}$ has components

$$E_x=\frac{1}{4\pi\varepsilon_0}\frac{(z'e)x}{(b^2+x^2)^{\frac{3}{2}}},\quad E_y=-\frac{1}{4\pi\varepsilon_0}\frac{(z'e)b}{(b^2+x^2)^{\frac{3}{2}}}.$$

Thus the change in momentum of the fast particle along its direction of motion is small, for if we approximate its motion by $x=vt$,

$$\Delta p_x=ze\int_{-\infty}^{\infty}E_x\,\mathrm{d}t\approx\left(\frac{zz'e^2}{4\pi\varepsilon_0}\right)\int_{-\infty}^{\infty}\frac{vt\,\mathrm{d}t}{(b^2+v^2t^2)^{\frac{3}{2}}}=0,$$

whereas the particle acquires transverse momentum $p_{\mathrm{T}}=\Delta p_y$ given by

$$p_{\mathrm{T}}=ze\int_{-\infty}^{\infty}E_y\,\mathrm{d}t=-\left(\frac{zz'e^2}{4\pi\varepsilon_0}\right)\int_{-\infty}^{\infty}\frac{b\,\mathrm{d}t}{(b^2+v^2t^2)^{\frac{3}{2}}}=-\left(\frac{zz'e^2}{4\pi\varepsilon_0}\right)\frac{2}{bv}. \tag{13.1}$$

(The integral is easily evaluated by the substitution $vt=b\tan\phi$.)

13.1 The approximate trajectory of a 'fast' particle passing a 'rest' particle.

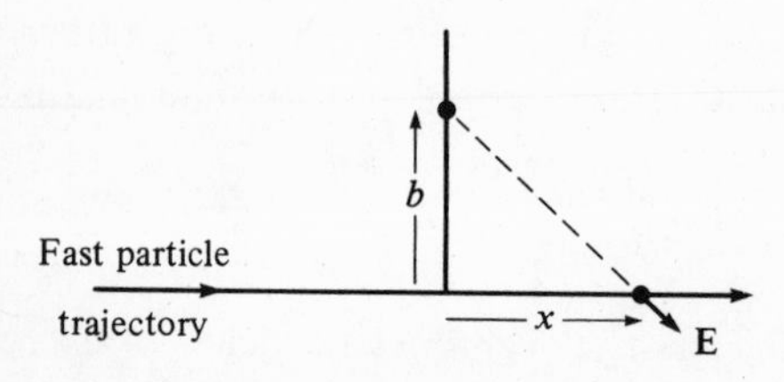

Since momentum is conserved overall, the 'rest' particle acquires momentum $(-p_T)$ and, assuming that it does not attain a relativistic velocity, gains kinetic energy $(p_T^2/2m_R)$. This energy must be lost by the fast particle:

$$\Delta E = -\frac{p_T^2}{2m_R} = -2\left(\frac{zz'e^2}{4\pi\varepsilon_0}\right)^2 \frac{1}{b^2v^2m_R}. \tag{13.2}$$

Note that ΔE does not depend on the mass of the fast particle, and that the calculation is valid for relativistic velocities of the fast particle.

In applying this result, the 'rest' particles are the atomic nuclei and atomic electrons of the gas. For an atomic nucleus of atomic number Z, $z' = Z$, and (except for hydrogen) $m_R \approx 2Zm_p$. For an electron $z' = -1$ and $m_R = m_e$. Using the formula (13.2), when a fast charged particle passes through a gas the ratio of the energy lost to the atomic electrons, to the energy lost to the atomic nuclei, is $\approx 2m_p/m_e \approx 4 \times 10^3$ (since each nucleus has Z electrons). Thus the energy lost to the nuclei is negligible compared with that lost to the electrons, and we shall only take the latter into account. (We are implicitly assuming that the velocity of the fast particle is large compared with the velocities of the atomic electrons in their orbits.)

If the gas is of mass density ρ, and consists of atoms of atomic number Z, atomic mass m_a, the number of electrons per unit volume is $(\rho/m_a)Z$. When the fast particle moves through a distance dx in the gas it passes, on average, $(\rho/m_a)Z\ 2\pi b\ db\ dx$ electrons with impact parameter between b and $b+db$, and the energy lost to these electrons is

$$d^2E = -4\pi\left(\frac{ze^2}{4\pi\varepsilon_0}\right)^2 \frac{\rho Z}{m_a} \frac{1}{v^2 m_e} \frac{db}{b} dx.$$

Integrating this expression over all impact parameters between b_{min} and b_{max}, the total rate of energy loss along the path, or *stopping power*, is

$$-\frac{dE}{dx} = 4\pi\left(\frac{ze^2}{4\pi\varepsilon_0}\right)^2 \frac{\rho Z}{m_a m_e} \frac{1}{v^2} L, \tag{13.3}$$

where $L = \log(b_{max}/b_{min})$.

Since $m_a \approx A$ atomic mass units, where A is the mass number of the atoms, we write this as

$$-\frac{dE}{dx} = D\left(\frac{Z}{A}\right)\rho\left(\frac{zc}{v}\right)^2 L, \tag{13.4}$$

where

$$D = 4\pi\left(\frac{e^2}{4\pi\varepsilon_0}\right)^2 \frac{1}{m_e(931.5\ \text{MeV})} = 0.307\ \text{MeV cm}^2/\text{g},$$

and the mass density ρ of the material is expressed in g/cm^3. (Note the units.)

We have introduced parameters b_{max} and b_{min}. Our formula (13.2) clearly breaks down for small b, since the energy transfer cannot be indefinitely large; it also breaks down at large b, since to ionise the atom the energy transfer cannot be indefinitely small. A quantum mechanical calculation by Bethe which holds for charged particles other than electrons and positrons gives equation (13.4) with

$$L=\left[\log\left(\frac{2\gamma^2 m_e v^2}{\langle I\rangle}\right)-\frac{v^2}{c^2}\right] \tag{13.5}$$

where $\langle I\rangle$ is a suitably-defined average ionisation energy over atomic electrons.

The form of (13.5) can be understood qualitatively from the following considerations. In quantum mechanics a particle is represented by a wave-packet, and for the classical treatment to be a good approximation the dimensions of the wave-packet Δx must surely be less than the impact parameter b. By the uncertainty principle, the minimum size of a wave-packet is $\hbar/p=\hbar/(\gamma mv)$. (For a particle of mass m moving relativistically we must include $\gamma=(1-v^2/c^2)^{-\frac{1}{2}}$.) In the centre-of-momentum frame of the two particles, the uncertainty in position is the same for both. If the fast particle is massive compared with an electron, then in the centre-of-momentum frame it is nearly at rest and the electron has velocity $\approx v$ and hence momentum $p\approx\gamma m_e v$. This suggests we should in this case take

$$b_{min}\approx\frac{\hbar}{\gamma m_e v}.$$

At large impact parameters the energy transfer is small, and we must recognise that the electrons are bound in atoms and have discrete energy levels, so that there will be a minimum energy of excitation, of the order of the ionisation energy I of the electron. In a 'collision time' τ_c the energy of a particle can be uncertain by $\Delta E=\hbar/\tau_c$, so that to transfer energy I requires $\tau_c<\hbar/I$. In our case the collision time $\tau_c\sim b/\gamma v$, where the factor γ comes from relativistic considerations, so we take

$$b_{max}=\frac{\hbar\gamma v}{I}.$$

We have then

$$L=\log\left(\frac{b_{max}}{b_{min}}\right)\approx\log\left(\frac{\gamma^2 m_e v^2}{\langle I\rangle}\right).$$

This expression is very similar to (13.5). $\langle I\rangle$ is usually treated as a parameter and determined by fitting the formula to experimental data. Though the formula is derived for gases, it is used for liquids and crystals by adjusting $\langle I \rangle$. For compounds 'Bragg's additivity rule' is found to hold quite well: the energy losses computed for each constituent separately may simply be added.

The energy loss increases as the particle slows down, due to the $1/v^2$ factor. (The logarithm is only a slowly varying function of its argument.) Hence the heaviest ionisation is found towards the end of the track. This is indeed the case, but the formula was derived for fast particles. At low velocities, of the order of the atomic electron velocities, it becomes invalid, and $(-\mathrm{d}E/\mathrm{d}x)$ decreases to zero as the particle comes to rest. Figure 13.2 plots the stopping power for protons in copper from experimentally determined data, together with the Bethe formula (13.5) with a fitted value of $\langle I \rangle$.

The energy-loss equation is of the form

$$-\frac{\mathrm{d}E}{\mathrm{d}x} = -\frac{\mathrm{d}T}{\mathrm{d}x} = z^2 \cdot (\text{function of } v/c)$$

13.2 The stopping power for protons and deuterons in copper. The theoretical curve is from equation (13.5) with $\langle I \rangle = 0.375$ keV. The experimental points are from Andersen, H. H. *et al.* (1966), *Phys. Rev.* **153**, 338.

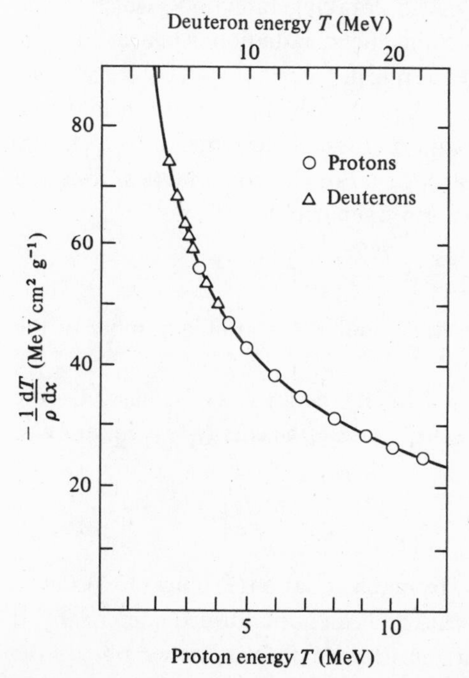

where T is the kinetic energy of the fast particle. In relativistic mechanics, T is given by

$$T=\frac{Mc^2}{\sqrt{(1-v^2/c^2)}}-Mc^2,$$

so that we can express (v/c) as a function of (T/Mc^2), and write also

$$-\frac{\mathrm{d}T}{\mathrm{d}x}=z^2F(T/Mc^2), \tag{13.6}$$

where M is the mass of the *fast* particle. We can use this result to scale the data for, say, protons, to apply to other particles. For example, the stopping power of copper for a 2.5 MeV proton is $\approx$ that for a 5 MeV deuteron, since $m_d \approx 2m_p$ and $z=1$ for both. The validity of this scaling is exemplified by the experimental data also shown in Fig. 13.2.

The result (13.5) holds for particles massive compared with an electron, and must be modified for fast electrons or positrons passing through matter. In particular, the momentum of either electron in the centre-of-mass system is $\gamma m_e v/\sqrt{[2(\gamma+1)]}$, so that $b_{\min}$ becomes $\hbar\sqrt{[2(\gamma+1)]}/\gamma m_e v$. There are other quantum corrections, and the expressions for electrons and positrons differ slightly. However, another energy-loss mechanism becomes significant for electrons and positrons at the higher energies in our range. This is *Bremsstrahlung* or, in English, 'braking radiation', which is the energy loss by emission of electromagnetic radiation when a charged particle accelerates and decelerates during its collisions with the constituent atoms of the matter it is passing through. We shall not treat this in detail. Bremsstrahlung is most significant in heavy elements, in which the Coulomb fields of the nuclei are strongest. For electrons and positrons, the ratio of energy loss rates is given approximately by

$$\frac{\text{Bremsstrahlung energy loss}}{\text{ionisation energy loss}} \approx \frac{T(Z+1.2)}{700},$$

where T is the kinetic energy in MeV, and Z the atomic number of the material.

The *range* $R(T_0)$ of a fast particle of initial kinetic energy T_0, mass M, is the mean distance it travels before it stops. Using the energy-loss equation in the form (13.6),

$$R(T_0)=\int_{T_0}^{0}\frac{\mathrm{d}T}{\mathrm{d}T/\mathrm{d}x}=\frac{1}{z^2}\int_0^{T_0}\frac{\mathrm{d}T}{F(T/(Mc^2))}=\frac{Mc^2}{z^2}\int_0^{T_0/Mc^2}\frac{\mathrm{d}u}{F(u)}.$$

Another useful scaling law follows from this equation (Problem 13.1). Given the expression for $F(T/(Mc^2))$ the integral can be performed numerically. If a constant 'mean' $\bar{L}$ is used in the formula for $\mathrm{d}T/\mathrm{d}x$ we obtain the approximate result

$$R(T) \approx \frac{(A/Z)}{D\rho z^2 \bar{L}} \left(\frac{T^2}{Mc^2 + T} \right). \tag{13.7}$$

Note the mass M appearing in the denominator. If Bremsstrahlung is negligible, a similar result holds for electrons, with $\bar{L}$ of the same order of magnitude for the same material. Thus it is clear that in a given material electrons and positrons travel a much greater distance on average than protons or other charged particles of the same kinetic energy. A positron may annihilate with an electron whilst still in motion, but even in lead, where electrons are abundant, the probability that it comes to rest before it annihilates is greater than 80%.

In connection with energy loss, it is often of interest to know the total amount of ionisation caused by the deposition of energy. The primary electrons knocked out of atoms may have sufficient energy to ionise further atoms. It is found experimentally that the total number of electron–ion pairs produced is closely proportional to the energy deposited, independently of the charge and velocity of the fast particle. The average energy deposited per electron–ion pair formed is, typically, 30–40 eV for gases, and 3–4 eV for semiconductors.

13.2 Multiple scattering of charged particles

We have assumed that the 'fast' particle travels along an approximately straight path, but it is of course deflected in collisions. The angle of deflection $\Delta\theta$ in a single collision is given by

$$\Delta\theta \approx \frac{p_{\mathrm{T}}}{p} = \left(\frac{zz'e^2}{4\pi\varepsilon_0} \right) \frac{2}{bpv},$$

using equation (13.1). This result agrees with the small-angle limit of the well-known Rutherford scattering formula.

If the deflections occur randomly, as in a gas, the mean square transverse momentum after several collisions (i) is given by

$$\overline{\mathbf{p}_{\mathrm{T}}^2} = \sum_i (\mathbf{p}_{\mathrm{T}}^i)^2,$$

treating the vectors $\mathbf{p}_{\mathrm{T}}^i$ as a random walk.

Since for the atomic nuclei $z' = Z$, and for the atomic electrons $z' = -1$, it is the *nuclei* rather than the electrons which are responsible for scattering the fast particle (except in hydrogen) and we can regard the electrons as simply screening the Coulomb field of the atomic nucleus at large impact parameters.

In a distance Δx, the fast particle passes $(\rho/m_a)\ 2\pi b\ \mathrm{d}b\ \Delta x$ nuclei with impact parameters between b and $b+\mathrm{d}b$, so that

$$\overline{\mathbf{p}_T^2} = \frac{2\pi\rho}{m_a}\left(\frac{zZe^2}{4\pi\varepsilon_0}\right)^2 \frac{4}{v^2}\int\frac{\mathrm{d}b}{b}\,\Delta x.$$

Again, we have to impose a maximum and minimum b. We should here take $b_{max} \approx$ an atomic dimension, beyond which the electrons screen out the field of the nucleus, and $b_{min} \approx$ a nuclear dimension, since for an energetic particle it is only when the impact parameter becomes comparable with the nuclear size that the Rutherford formula gives large-angle scattering, and our approximate expression breaks down. Such large-angle scatterings, though historically important (§4.1), are rare. At even smaller distances the Coulomb field of the nucleus is modified and nuclear interactions might occur.

Hence we have, roughly, $\int \mathrm{d}b/b \approx \log(\text{Å}/\text{fm}) \approx 10$, and the mean squared deflection in a path length Δx is given by

$$\overline{\theta^2} \approx \frac{\overline{\mathbf{p}_T^2}}{p^2} \approx (3\ \mathrm{cm}^2\ \mathrm{g}^{-1}\ \mathrm{MeV}^2)\frac{\rho\Delta x}{(pv)^2}\frac{Z^2z^2}{A}.$$

Thus the effect of multiple scattering increases rapidly with the atomic number Z of the material. For heavy particles, $pv = mv^2 = 2T$, and for energetic electrons, $pv \approx pc \approx T$. Hence the mean square deflection per unit length is not very sensitive to the mass of the particle, at given kinetic energy. However, the effects of multiple scattering become more evident for electrons and positrons because of their longer path length, and the concept of a well-defined linear range is not applicable to these particles.

13.3 Energetic photons

An energetic charged particle loses its energy in small fractions as it passes through matter, and for a given initial energy it travels a fairly well-defined distance before it comes to rest. When an energetic photon interacts with an atom, it is totally absorbed or scattered. The intensity of an initially collimated beam of photons is thereby attenuated, while the individual photons left in the beam are unscathed. It is useful to define an *attenuation coefficient*, in terms of the total cross-section, $\sigma_{tot}(E)$, for a photon of energy $E = \hbar\omega$ to interact with an atom. We consider a thin section of the material, of area S, thickness $\mathrm{d}x$, normal to the direction of the photons. If the material is of density ρ and made up of atoms of mass m_a, the section will contain $(\rho/m_a)S\ \mathrm{d}x$ atoms. For sufficiently small $\mathrm{d}x$, multiple scatterings can be neglected. If n incident photons impinge at random on the slab, then from the definition of cross-section (Appendix A) the number of photons dn lost from the beam is given by (Fig. 13.3):

$$\frac{dn}{n} = -\frac{\text{area covered by cross-sections}}{\text{total area}}$$

$$= -\frac{(\rho/m_a)S\,dx\sigma_{tot}}{S} = -(\rho/m_a)\sigma_{tot}\,dx.$$

Integrating this equation yields

$$n(x) = n(0)e^{-\mu x},$$

where $\mu = \rho\sigma_{tot}/m_a$ is called the linear attenuation coefficient. It is usual to give data in terms of the *mass attenuation coefficient* $(\mu/\rho) = \sigma_{tot}/m_a$, as a function of photon energy. The attenuation coefficient for compounds can be calculated to within a few per cent by assuming

$$\mu = \sum_i \rho_i(\sigma_{tot}^i/m_a^i),$$

where i labels the ith constituent.

In the range of photon energies from 1 keV to 10 MeV there are three main contributions to σ_{tot}: photo-electric absorption in atoms, the Compton effect, and pair production. The relative importance of these contributions varies with energy (Fig. 13.4). These processes do not interfere with each other and we can take

$$\sigma_{tot} = \sigma_a + Z\sigma_C + \sigma_p.$$

At low energies the *photo-electric effect* is dominant. In this process the photon is absorbed completely by the atom, and an atomic electron is raised to a higher unoccupied bound state or an unbound state. In Fig. 13.4 it will be seen that the cross-section for photo-electric absorption σ_a rises sharply at the energies corresponding to the onset of ionisation of the L and

13.3 Effective total cross-section presented to photons incident normally on a slab of area S, thickness dx.

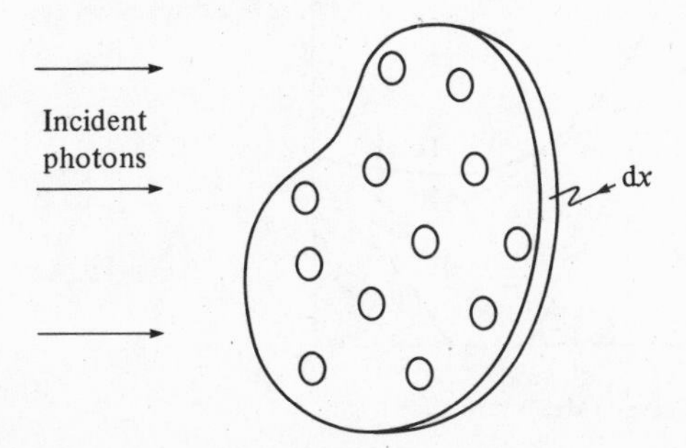

K shell electrons. These energies are higher in heavier elements in which the core electrons are more tightly bound (Fig. 13.5).

At energies above the K shell absorption edge, the photo-electric absorption falls off and the scattering of photons by electrons takes over as the main contribution to the attenuation. This is *Compton scattering*, and some of the photon energy goes into the recoil energy of the electron. At these energies the atomic binding of the electrons may be neglected and they may be treated as free. The momentum–energy conservation calculation is elementary and well known. The calculation of the Compton scattering cross-section σ_C, like those of σ_a and σ_p, is a well-understood calculation in quantum electrodynamics, but the order of magnitude of σ_C can be found from a simple classical calculation which is valid for low energies ($\hbar\omega \ll m_e c^2$) at which the electron recoil is negligible.

In a classical picture, the electron vibrates with the frequency of the incident electromagnetic wave and emits a secondary wave at the same frequency. The result of the classical calculation (Problem 13.7) is an

13.4 Contributions to the mass attenuation coefficient for photons in lead. (Data for this figure and Fig. 13.5 from Review of Particle Properties (1983), *Rev. Mod. Phys.* **56**, S1.)

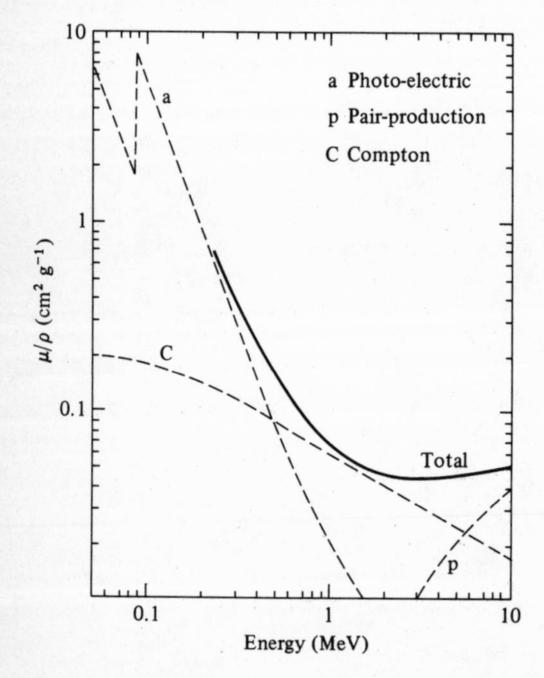

effective cross-section σ_T where

$$\sigma_T = \frac{8\pi}{3}\left(\frac{e^2}{4\pi\varepsilon_0}\right)^2\left(\frac{1}{m_e c^2}\right)^2$$
$$= 0.665 \times 10^{-28}\ \text{m}^2 = 0.665\ \text{b}.$$

This limiting value, which does not depend on frequency, is known as the *Thomson scattering* cross-section. At higher energies ($\hbar\omega \gtrsim m_e c^2$) quantum effects become important, and the Compton-scattering cross-section falls below this value. The corresponding Compton scattering from the atomic nuclei is reduced by many orders of magnitude because of the (mass)^2 factor in the denominator.

Since there are Z electrons for each atom, the attenuation due to

13.5 The mass attenuation coefficient for photons in some elements spanning the periodic table. Note that both the horizontal and vertical scales are logarithmic.

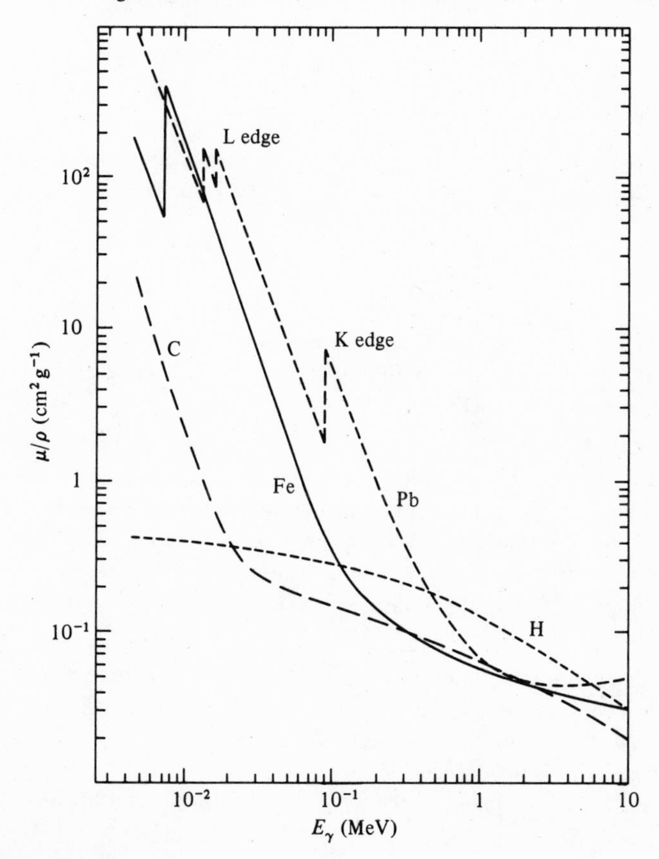

Compton scattering is given by

$$\mu_C = (\rho/m_a) Z \sigma_C \approx \frac{\rho}{1\ \text{amu}} \left(\frac{Z}{A}\right) \sigma_C.$$

For elements other than hydrogen $(Z/A) \approx \frac{1}{2}$, so that their plots of (μ/ρ) approximately coincide in the energy range where Compton scattering dominates, as Fig. 13.5 shows. In this figure the value of (μ/ρ) for hydrogen at low energies is close to $(\sigma_T/m_H) = 0.397\ \text{cm}^2\ \text{g}^{-1}$.

At photon energies $\hbar\omega > 2m_e c^2 \approx 1.02$ MeV *pair production* becomes possible (§2.3). This is the process

$$\gamma + (\text{nucleus}) \rightarrow e^+ + e^- + (\text{nucleus}),$$

which can occur most readily in the Coulomb field of a heavy nucleus. The cross-section σ_p increases with energy, and eventually pair-production dominates over other processes. It can be regarded as the inverse process to Bremsstrahlung, and the cross-section σ_p increases with Z similarly. This is why the turn-up in the curves of Fig. 13.5 is most pronounced in the case of lead.

Problems

13.1 Show that if $R_p(T)$ is the range of a proton of kinetic energy T, the range $R_M(T_M)$ of a charged particle of mass M, kinetic energy T_M, and charge ze is given by

$$R_M(T_M) = \frac{M}{z^2 m_p} R_p(m_p T_M / M).$$

13.2 If L in equation (13.4) is replaced by a constant $\bar{L}$, show that the integral for the range of an ionising particle can be evaluated to give the approximate result (13.7).

13.3 For 'back-of-envelope' calculations, a useful estimate of the mean ionisation energy $\langle I \rangle$ for an atom of atomic number Z is $\langle I \rangle = 12Z$ eV. Show that for α-particles of kinetic energy 2 MeV in aluminium the L of equation (13.5) ≈ 2; for electrons of kinetic energy 2 MeV in aluminium, $L \approx 10$. Use these values to estimate the range of 5 MeV α-particles and of 5 MeV electrons in aluminium (mass density 2.7 g cm^{-3}).

13.4 Show that for a non-relativistic particle of mass M, velocity v, $(\mathrm{d}E/\mathrm{d}x) = M(\mathrm{d}v/\mathrm{d}t)$. Replacing L by a constant $\bar{L}$ in equation (13.4), show that the time for a non-relativistic particle with initial velocity v_0 to come to rest is $(4/3)(\text{range})/v_0$.

Estimate the time taken by the α-particle of Problem 13.3 to come to rest.

13.5 In a neutron detector of the type described in Problem 8.2, estimate roughly the number of ion pairs produced in the helium gas per neutron interaction and the distance over which the ionisation is deposited.

13.6 50 keV X-rays are in common use in dentistry. Estimate the thickness of lead sheet (density 11.4 g cm^{-3}) that will absorb 99.9% of such radiation at normal incidence.

13.7 Larmor's formula for the power P radiated from a non-relativistic particle of charge e and acceleration a is

$$P=\frac{2}{3}\left(\frac{e^2}{4\pi\varepsilon_0}\right)\frac{a^2}{c^3}.$$

Show that classically an electron in an electric field $\mathbf{E}=\mathbf{E}_0\cos\omega t$ will radiate energy at a mean rate

$$P=\frac{1}{3}\frac{e^2|\mathbf{E}_0|^2}{m_e^2c^3}\left(\frac{e^2}{4\pi\varepsilon_0}\right).$$

The incident energy flux in a plane electromagnetic wave is $c\varepsilon_0|\mathbf{E}_0|^2/2$ (cf. § 12.9). Hence obtain the Thomson scattering formula.

Appendix A
Cross-sections

We begin this appendix by considering neutron cross-sections. There is some simplification in the case of neutrons, since they are electrically neutral and do not interact through the long-range Coulomb force. To a good approximation they can be considered to interact only through the short-range nuclear force. The concepts developed for neutrons may be applied almost immediately to other electrically-neutral particles, such as photons. We then turn to the case of charged particles.

A.1 Neutron and photon cross-sections

We consider a neutron approaching from a distance a nucleus which is at rest. (Any interactions between the neutron and the atomic electrons will be neglected.) We suppose that, if the nucleus were not present, the probability of the neutron passing anywhere through a circle of radius a, centred on the nucleus and perpendicular to the direction of the neutron's motion, would be uniform (Fig. A1), i.e. the probability of it passing through an area δA would be $\delta A/(\pi a^2)$. We can think of the neutron as a classical particle, or better, as a quantum-mechanical wave-packet. The radius a must be large compared with both the size of the wave-packet and the size of the nucleus. With the nucleus present, an interaction can take place, for example scattering, induced fission, or radiative capture. It is found that, provided a is large enough, the probability of an interaction is inversely proportional to the area πa^2, i.e.

$$\text{probability of interaction} = \frac{\sigma_{\text{tot}}}{\pi a^2}. \tag{A.1}$$

The constant of proportionality introduced here, σ_{tot}, is called the *total cross-section.* Clearly σ_{tot} has the dimensions of area. It can be regarded as the effective area presented to the neutron by the nucleus, but it must be realised that the cross-section is a *joint* property of the neutron and nucleus, and for a given nucleus it is a sensitive function of neutron energy. The probability of interaction is a quantum-mechanical property: σ_{tot} can be very much larger than the geometrical cross-section of the nucleus.

The system of a moving particle incident on a nucleus at rest is called the *laboratory system.* With fixed targets, it is the situation most easy to simulate in the laboratory. It is also useful to consider interactions in the frame of reference in which the nucleus has momentum equal in magnitude but opposite in direction to the neutron. It is clear from the definition that the total cross-section is the same viewed in this '*centre-of-mass*' or 'centre-of-momentum' system as in the laboratory system.

There will usually be several possible *reaction channels*, i.e. types of interaction that can occur. Examples are:

Elastic scattering: the incoming neutron changes direction but, in the centre-of-mass system, loses no energy.

Inelastic scattering: the incoming neutron changes direction and, even in the centre-of-mass system, loses energy in exciting the nucleus.

Radiative capture: the incoming neutron is captured by the nucleus. The resulting nucleus is formed in an excited state, which eventually decays by photon emission.

Given that a reaction occurs, each reaction channel 'i' has a definite probability p_i, where

$$\sum_i p_i = 1.$$

The *partial cross-section*, σ_i for the ith channel, is defined to be $\sigma_i = p_i \sigma_{\text{tot}}$, and may be regarded as the effective area presented by the target nucleus to the neutron for that particular reaction. We have

$$\sigma_{\text{tot}} = \sum_i \sigma_i.$$

Photon cross-sections can be defined in direct analogy with neutron cross-sections, but since photons interact with atomic electrons as well as with nuclei it is more appropriate to consider the target to be an atom.

A1 Neutron incident on area centred on nucleus at rest.

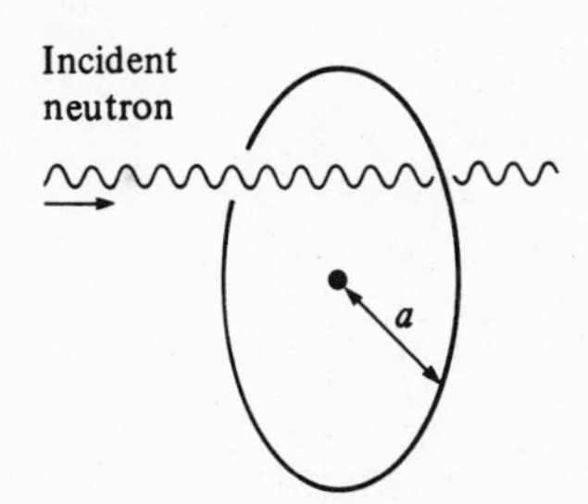

Various contributions to the total cross-section for a photon to interact with an atom are discussed in Chapter 13.

A.2 Differential cross-sections

In considering a particular reaction channel it is often useful to subdivide it further. For example, in the elastic scattering of neutrons it can be of interest to know the probability distribution of the angle at which the neutron emerges from the interaction. Given that an elastic scattering occurs, if $p_e(\theta)\,d\Omega$ is the probability that the neutron is scattered into a small solid angle $d\Omega = 2\pi \sin\theta\, d\theta$, at a polar angle θ with respect to its incident direction, we write (Fig. A2)

$$p_e(\theta)\,d\Omega = \frac{1}{\sigma_e}\left(\frac{d\sigma_e}{d\Omega}\right) d\Omega. \tag{A.2}$$

This defines the *elastic differential cross-section* $d\sigma_{el}/d\Omega$, and

$$\int \frac{d\sigma_e}{d\Omega}\, d\Omega = \sigma_e. \tag{A.3}$$

Differential cross-sections are usually measured in the laboratory with respect to a fixed target. In the centre-of-mass frame the direction of a scattered neutron, and hence the angular dependence of the cross-section, will be different. The kinematic transformation between the frames is straightforward, and experimental data is often presented in the centre-of-mass frame to facilitate comparison with theory.

A.3 Reaction rates

Consider a broad collimated beam of mono-energetic neutrons. Let ρ_n be the number density of neutrons in the beam, and v the neutron velocity. The *neutron flux*, i.e. the number of neutrons crossing a unit area normal to the beam per unit time, is $\rho_n v$. Hence in time dt, the number of neutrons passing through a circle of radius a centred on a nucleus is $\rho_n v\, dt\, \cdot$

A2 Geometry of elastic scattering from a fixed target.

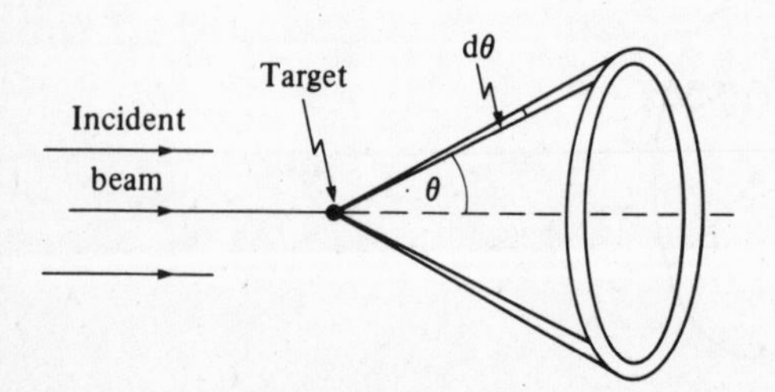

πa^2. From (A.1), the probability of a reaction with the nucleus taking place in the time interval $t, t+\mathrm{d}t$, given the nucleus is in its ground state at time t, is $\rho_\mathrm{n} v \sigma_\mathrm{tot}\, \mathrm{d}t$. Thus the *reaction rate* per nucleus is $\rho_\mathrm{n} v \sigma_\mathrm{tot}$ or

reaction rate = flux × cross-section.

We may also consider a single neutron, moving with velocity v in a random array of nuclei of number density ρ_nuc. By the same argument (in the frame in which the neutron is at rest and the nuclei are regarded as a beam) the reaction rate is $\rho_\mathrm{nuc} v \sigma_\mathrm{tot}$. Given that the neutron exists at $t=0$, the mean time τ before an interaction takes place is, therefore, $\tau = (\rho_\mathrm{nuc} v \sigma_\mathrm{tot})^{-1}$, and the *mean free path* l, the distance it travels in this time, is $l = v\tau = 1/(\rho_\mathrm{nuc}\sigma_\mathrm{tot})$. (We have assumed that τ is very much shorter than the quarter of an hour intrinsic mean life of the neutron.)

A.4 Charged particle cross-sections: Rutherford scattering

The difficulties that arise in charged particle scattering stem from the long range of the Coulomb force. In § 13.2 it is shown that a charged particle, say a proton, passing a fixed target, say a nucleus of charge Ze, is deflected through an angle θ given approximately by

$$\theta = \left(\frac{Ze^2}{4\pi\varepsilon_0}\right)\frac{2}{bpv} \tag{A.4}$$

where p is the momentum and v the velocity of the proton, and b is the impact parameter, the distance at which the proton would pass the nucleus if there were no interaction (Fig. 13.1).

Impact parameters between b and $b+\mathrm{d}b$ correspond to scattering angles between θ and $\theta - \mathrm{d}\theta$ where

$$\mathrm{d}\theta = \frac{\mathrm{d}b}{b^2}\left(\frac{2Ze^2}{4\pi\varepsilon_0 pv}\right).$$

The effective area presented to the proton which corresponds to this range $\mathrm{d}b$ of impact parameters is $2\pi b\, \mathrm{d}b$, and we can interpret this as a contribution to the elastic scattering cross-section,

$$\begin{aligned}\mathrm{d}\sigma_\mathrm{e} &= 2\pi b\, \mathrm{d}b \\ &= 2\pi\left(\frac{2Ze^2}{4\pi\varepsilon_0 pv}\right)^2 \frac{\mathrm{d}\theta}{\theta^3}, \quad \text{using (A.4).}\end{aligned}$$

The differential scattering cross-section for small angles is therefore

$$\frac{\mathrm{d}\sigma_\mathrm{e}}{\mathrm{d}\Omega} = \frac{1}{2\pi \sin\theta}\frac{\mathrm{d}\sigma_\mathrm{e}}{\mathrm{d}\theta} = \left(\frac{2Ze^2}{4\pi\varepsilon_0 pv}\right)^2 \frac{1}{\theta^4}$$

where we have replaced $\sin\theta$ by θ. This is the small-angle limit of the famous Rutherford scattering formula. The same expression is obtained from a quantum mechanical calculation.

The differential cross-section becomes very large when θ is very small, and the total elastic cross-section, defined by the integral (A.3), and hence the total cross-section, is infinite. Physically this is because the Coulomb force is still felt by the proton no matter how large the impact parameter. In practice this formally infinite result is not a serious difficulty, since there is always a limit to the experimentalist's ability to measure small-angle scattering, and if one is interested only in elastic scattering through angles greater than some minimum angle the cross-section is finite.

At large impact parameters the Coulomb force is weak, and can only give rise to small-angle elastic scattering. Thus the cross-sections for other possible processes are all finite.

Appendix B
Density of states

Consider a particle moving freely inside a cubic box of side L, volume $V=L^3$. We take the potential to be zero inside the box, and represent the walls by infinite potential barriers. The Schrödinger equation for the particle,

$$-\frac{\hbar^2}{2m}\nabla^2\psi=E\psi, \tag{B.1}$$

is separable in (x, y, z) coordinates, and the solutions must vanish at the walls which we can take to be the planes $x=0$ and $x=L$, $y=0$ and $y=L$, $z=0$ and $z=L$. These solutions are easily seen to be standing waves of the form

$$\psi(x, y, z)=(\text{constant})\sin(k_x x)\sin(k_y y)\sin(k_z z), \tag{B.2}$$

provided that we choose $\mathbf{k}=(k_x, k_y, k_z)$ from the values

$$k_x=\frac{n_x\pi}{L},\quad n_x=1, 2, 3, \ldots;\quad k_y=\frac{n_y\pi}{L},\quad n_y=1, 2, 3, \ldots;$$

$$k_z=\frac{n_z\pi}{L},\quad n_z=1, 2, 3, \ldots$$

to satisfy the boundary conditions. Negative integer values of n_x, n_y, n_z do not give new states, since they merely change the sign of the wave-function, and such a phase factor has no physical significance.

Thus the allowed values of $\mathbf{k}$ form a cubic lattice of points in the $(+,+,+)$ quadrant of '$\mathbf{k}$-space'. Each eigenstate (B.2) corresponds to one point of the lattice, and counting states is equivalent to counting lattice points. The spacing between these lattice points is (π/L), so that the number of points per unit 'volume' in $\mathbf{k}$-space is $(L/\pi)^3$. The number of lattice points with $k\ (=|\mathbf{k}|)$ less than some fixed value k_0 is the number enclosed within the quadrant of the sphere centred at the origin and of radius k_0. This number must of course be an integer, but for large values of k_0 it will be approximately given by:

(volume of quadrant of sphere) × (density of lattice points)

$$=\frac{1}{8}\frac{4\pi k_0^3}{3}\left(\frac{L}{\pi}\right)^3=\frac{V}{(2\pi)^3}\frac{4\pi k_0^3}{3}. \tag{B.3}$$

The number of points with k lying in the range $k_0<k<k_0+\mathrm{d}k_0$ is the differential of (B.3);

$$\frac{V}{(2\pi)^3}4\pi k_0^2\,\mathrm{d}k_0. \tag{B.4}$$

We will consider the case of a spin $\frac{1}{2}$ fermion (for example, an electron or a nucleon). Then two states ('spin-up' and 'spin-down') can be assigned to each $\mathbf{k}$ value, and from (B.3) the number of states $\mathcal{N}_0$ with $k<k_0$ is

$$\mathcal{N}_0=2\frac{V}{(2\pi)^3}\frac{4\pi k_0^3}{3},\quad \text{or}\quad k_0^3=3\pi^2\frac{\mathcal{N}_0}{V}. \tag{B.5}$$

For the non-relativistic Schrödinger equation (B.1), the energy E of a particle in a state of specified (n_x, n_y, n_z) and either spin, is related to k by

$$E=\frac{\hbar^2}{2m}(k_x^2+k_y^2+k_z^2)=\frac{\hbar^2}{2m}k^2. \tag{B.6}$$

The *integrated density of states* $\mathcal{N}(E)$ is defined as the number of states with energy less than E. From (B.6) $k=(2mE/\hbar^2)^{\frac{1}{2}}$; hence using (B.5) we have

$$\mathcal{N}(E)=\frac{V}{3\pi^2}\left(\frac{2mE}{\hbar^2}\right)^{\frac{3}{2}}. \tag{B.7}$$

The *density of states* $n(E)=\mathrm{d}\mathcal{N}/\mathrm{d}E$, so that $n(E)\,\mathrm{d}E$ is the number of states with energy between E and $E+\mathrm{d}E$:

$$n(E)=\frac{\mathrm{d}\mathcal{N}}{\mathrm{d}E}=\frac{V}{2\pi^2}\left(\frac{2m}{\hbar^2}\right)^{\frac{3}{2}}E^{\frac{1}{2}}. \tag{B.8}$$

If the spin factor is omitted,

$$n(E)=\frac{V}{4\pi^2}\left(\frac{2m}{\hbar^2}\right)^{\frac{3}{2}}E^{\frac{1}{2}}. \tag{B.9}$$

In scattering problems, it is convenient to consider a large volume L^3 and impose 'periodic boundary conditions' on the wave-functions:

$$\psi(x+L, y, z)=\psi(x, y, z),$$
$$\psi(x, y+L, z)=\psi(x, y, z),$$
$$\psi(x, y, z+L)=\psi(x, y, z).$$

Instead of the standing waves (B.2), the solutions of the wave equation consistent with the boundary conditions are the travelling waves

$$e^{i\mathbf{k}\cdot\mathbf{r}}=e^{ik_x x}e^{ik_y y}e^{ik_z z},$$

where, to satisfy the periodicity conditions, we must now take

$$k_x=\frac{2n_x\pi}{L}, \quad n_x=0, \pm 1, \pm 2, \ldots, \quad \text{etc.}$$

The density of points in **k**-space becomes $(L/2\pi)^3$. However, permutations of sign $(\pm k_x, \pm k_y, \pm k_z)$ now correspond to different states (travelling waves in different directions), and the lattice points corresponding to distinct states with $|\mathbf{k}|<k_0$ fill the whole sphere of radius k_0 in **k**-space. We thus arrive again at the results (B.3) and (B.5); (B.7) and (B.8), which hold for non-relativistic spin $\frac{1}{2}$ fermions, are also still valid.

In fact, in the limit when the linear dimensions of the box become large compared with the de Broglie wavelength of the particle at energy E, the result for the density of states at energy E becomes independent both of the boundary conditions imposed and of the shape of the box, provided this remains simple. The integrated density of states in a sphere is illustrated in Fig. 5.2.

Problems

B.1(*a*) For a single particle in a large volume V, show that the number of allowed **k**-values in a small volume $\mathrm{d}^3\mathbf{k}=dk_x\, dk_y\, dk_z$ of **k**-space is

$$\frac{V}{(2\pi)^3}\,\mathrm{d}^3\mathbf{k}.$$

(*b*) Show that, for two particles (1) and (2), the wave vector associated with the centre-of-mass motion is $\mathbf{K}=\mathbf{k}_1+\mathbf{k}_2$ and with the relative motion is $\mathbf{k}=(m_1\mathbf{k}_2-m_2\mathbf{k}_1)/(m_1+m_2)$. Hence show

$$\mathrm{d}^3\mathbf{K}\,\mathrm{d}^3\mathbf{k}=\mathrm{d}^3\mathbf{k}_1\,\mathrm{d}^3\mathbf{k}_2$$

and that the number of $(\mathbf{K}, \mathbf{k})$ values with K in the range $K, K+\mathrm{d}K$ and k in the range $k, k+\mathrm{d}k$ is $(V^2/4\pi^4)K^2\,\mathrm{d}K\,k^2\,\mathrm{d}k$ if the particles are distinguishable, but $(V^2/8\pi^4)K^2\,\mathrm{d}K\,k^2\,\mathrm{d}k$ if the particles are identical.

Appendix C
Angular momentum

Students are referred to texts on quantum mechanics for the derivations of the results summarised in this appendix, which is intended as no more than an aide-mémoire.

C.1 Orbital angular momentum

In the shell model of both atomic and nuclear physics the single-particle Schrödinger equation, neglecting effects of the intrinsic spin of the particle, is of the form

$$H\psi=\left(-\frac{\hbar^2}{2M}\nabla^2+V(r)\right)\psi(\mathbf{r})=E\psi(\mathbf{r}) \tag{C.1}$$

where the potential energy $V(r)$ is spherically symmetric, a function of the radial coordinate r only. Because of spherical symmetry, the operator ∇^2 is most useful in spherical polar coordinates (r, θ, ϕ), in which the Schrödinger equation takes the form

$$-\frac{\hbar^2}{2M}\frac{1}{r}\frac{\partial^2}{\partial r^2}(r\psi)+\left(\frac{\mathbf{L}^2}{2Mr^2}+V(r)\right)\psi=E\psi \tag{C.2}$$

where $\mathbf{L}^2=L_x^2+L_y^2+L_z^2$ and $\mathbf{L}$ is the *orbital angular momentum* operator,

$$\mathbf{L}=\mathbf{r}\times\mathbf{p}=\mathbf{r}\times(-i\hbar\,\nabla).$$

$\mathbf{L}$ acts only on the angular coordinates (θ, ϕ). For example,

$$L_z=-i\hbar\left(x\frac{\partial}{\partial y}-y\frac{\partial}{\partial x}\right)=-i\hbar\frac{\partial}{\partial\phi}.$$

From the definition of $\mathbf{L}$, it is not difficult to obtain the commutation relations

$$[L_x, L_y]=i\hbar L_z,\quad [L_y, L_z]=i\hbar L_x,\quad [L_z, L_x]=i\hbar L_y;$$
$$[\mathbf{L}^2, L_x]=[\mathbf{L}^2, L_y]=[\mathbf{L}^2, L_z]=0. \tag{C.3}$$

Because L_x, L_y, L_z do not commute, it is not generally possible for a wave-function to be simultaneously an eigenstate of any two of them, but it is always possible to construct simultaneous eigenstates of $\mathbf{L}^2$ and any one of L_x, L_y, L_z. It is conventional to choose $\mathbf{L}^2$ and L_z.

The simultaneous eigenstates of $\mathbf{L}^2$ and L_z are denoted by $Y_{lm}(\theta, \phi)$, where

$$\begin{aligned} \mathbf{L}^2 Y_{lm} &= l(l+1)\hbar^2 Y_{lm} \\ L_z Y_{lm} &= m\hbar\, Y_{lm}. \end{aligned} \tag{C.4}$$

The allowed values of l are the integers $l=0, 1, 2, 3, \ldots$ and, for a given l, m takes one of the $(2l+1)$ values $-l, -l+1, \ldots, l-1, l$. The functions $Y_{lm}(\theta, \phi)$ are well-known *spherical harmonics*, and are normalised so that

$$\begin{aligned} \int Y^*_{l'm'} Y_{lm}\, \mathrm{d}\Omega &= \int_0^\pi \mathrm{d}\theta \sin\theta \int_0^{2\pi} \mathrm{d}\phi Y^*_{l'm'}(\theta, \phi) Y_{lm}(\theta, \phi) \\ &= \delta_{ll'}\, \delta_{mm'}. \end{aligned}$$

For example, $Y_{00} = 1/\sqrt{4\pi}$; a state of zero orbital angular momentum is spherically symmetric. Also

$$\begin{aligned} Y_{11} &= -\sqrt{\frac{3}{8\pi}}\frac{x+iy}{r} = -\sqrt{\frac{3}{8\pi}} \sin\theta\, e^{i\phi} \\ Y_{10} &= \sqrt{\frac{3}{4\pi}}\frac{z}{r} = \sqrt{\frac{3}{4\pi}} \cos\theta \\ Y_{1-1} &= \sqrt{\frac{3}{8\pi}}\frac{x-iy}{r} = \sqrt{\frac{3}{8\pi}} \sin\theta\, e^{-i\phi}. \end{aligned}$$

Note that these states with $l=1$ can be formed from the components of the unit vector $(x/r,\ y/r,\ z/r)$.

These examples illustrate a general rule: the parity of a state of given l is $(-1)^l$.

From (C.2), the eigenfunctions of the Schrödinger equation are of the form

$$\psi_{nlm} = u_{nl}(r) Y_{lm}(\theta, \phi) \tag{C.5}$$

where u_{nl} satisfies the ordinary differential equation

$$-\frac{\hbar^2}{2M}\frac{1}{r}\frac{\mathrm{d}^2}{\mathrm{d}r^2}(ru_{nl}) + \left(\frac{\hbar^2(l+1)}{2Mr^2} + V(r)\right) u_{nl}(r) = E_{nl} u_{nl}(r).$$

There are several examples of potentials $V(r)$ for which the radial functions $u_{nl}(r)$ are elementary, and many others for which the numerical solutions are easy to programme on computers.

Note that the energy eigenstates (C.5) are also eigenstates of $\mathbf{L}^2$ and L_z. This is only possible because of the spherical symmetry of $V(r)$, which allows $\mathbf{L}^2$ and L_z (which act on θ and ϕ only) to commute with the energy operator.

C.2 Intrinsic angular momentum

A particle may have an *intrinsic angular momentum* or *spin* **s**, satisfying the same commutation relations as (C.3). The eigenvalues of $\mathbf{s}^2$ are $s(s+1)\hbar^2$, and m_s can take $(2s+1)$ values from $-s$ to $+s$. In the case of orbital angular momentum treated above, l must be a positive integer. This condition stems from the single-valuedness of the wave-function in space. The quantum number s is not subject to this restriction, since the coordinates on which **s** acts are internal to the particle, and we require only that $(2s+1)$ should be an integer. Thus we may have $s=\frac{1}{2}$, as is the case with leptons and nucleons. For $s=\frac{1}{2}$ there are two eigenstates, corresponding to $m_s=+\frac{1}{2}, m_s=-\frac{1}{2}$. We may denote these by $|+\frac{1}{2}\rangle$ and $|-\frac{1}{2}\rangle$. A general wave-function for the spin $\frac{1}{2}$ fermion is a superposition of 'spin-up' and 'spin-down' states of the form

$$\psi(\mathbf{r}, m_s)=\psi_+(\mathbf{r})|+\tfrac{1}{2}\rangle+\psi_-(\mathbf{r})|-\tfrac{1}{2}\rangle.$$

The two independent spin states $|+\frac{1}{2}\rangle$, $|-\frac{1}{2}\rangle$ may be represented by column vectors

$$|+\tfrac{1}{2}\rangle=\begin{pmatrix}1\\0\end{pmatrix}, \quad |-\tfrac{1}{2}\rangle=\begin{pmatrix}0\\1\end{pmatrix}.$$

s_x, s_y, s_z are then represented by 2×2 matrices. It is convenient to take out a factor $(\hbar/2)$ and write $\mathbf{s}=(\hbar/2)\boldsymbol{\sigma}=(\hbar/2)(\sigma_x, \sigma_y, \sigma_z)$. It is easy to verify that the commutation relations are satisfied using the *Pauli matrices*:

$$\sigma_x=\begin{pmatrix}0 & 1\\1 & 0\end{pmatrix}, \quad \sigma_y=\begin{pmatrix}0 & -i\\i & 0\end{pmatrix}, \quad \sigma_z=\begin{pmatrix}1 & 0\\0 & -1\end{pmatrix}.$$

$|+\frac{1}{2}\rangle$ and $|-\frac{1}{2}\rangle$ are eigenvectors of σ_z with eigenvalues $+1$ and -1.

C.3 Addition of angular momenta

The total angular momentum of a spin $\frac{1}{2}$ fermion is the sum of its orbital and intrinsic angular momenta:

$$\mathbf{J}=\mathbf{L}+\mathbf{s}.$$

It is easy to see that **J** satisfies commutation relations similar to (C.3), and also $[\mathbf{J}, \mathbf{L}^2]=0$, $[\mathbf{J}, \mathbf{s}^2]=0$. It is therefore possible to find states which are simultaneous eigenstates of $\mathbf{L}^2, \mathbf{s}^2, \mathbf{J}^2$ and J_z, specified by quantum numbers (l, s, j, j_z). These states have parity $(-1)^l$.

For a given value of l and $s=\frac{1}{2}$ there are $2\times(2l+1)=4l+2$ independent states, $Y_{lm}|\pm\frac{1}{2}\rangle$. We seek the linear combinations of these which are the eigenstates of $\mathbf{J}^2$ and J_z. Since $J_z=L_z+s_z$, the maximum value of j_z is $l+\frac{1}{2}$, corresponding to the state $Y_{ll}|+\frac{1}{2}\rangle$. This must also be the maximum value of

j, and the state must also be an eigenstate of $\mathbf{J}^2$ corresponding to $j=l+\frac{1}{2}$, $j_z=l+\frac{1}{2}$.

There are two independent states giving $j_z=l-\frac{1}{2}$, i.e. $Y_{l,l-1}|+\frac{1}{2}\rangle$, $Y_{l,l}|-\frac{1}{2}\rangle$. From these we must be able to construct the state corresponding to $j=l+\frac{1}{2}, j_z=l-\frac{1}{2}$. Another independent state can also be constructed; this clearly must correspond to $j=l-\frac{1}{2}, j_z=l-\frac{1}{2}$.

The value $j=l+\frac{1}{2}$ gives $2(l+\frac{1}{2})+1=2l+2$ states; the value $j=l-\frac{1}{2}$ gives $2(l-\frac{1}{2})+1=2l$ states. Altogether, we have $(4l+2)$ independent states, corresponding to the values $j=l+\frac{1}{2}, j=l-\frac{1}{2}$, and there can be no more allowed values of j. We can think of the intrinsic spin $\mathbf{s}$ of the particle as either aligned or anti-aligned with the orbital angular momentum vector $\mathbf{L}$, in so far as the uncertainty principle allows.

More generally, for two particles, or two systems, with angular momenta $\mathbf{J}_1$ and $\mathbf{J}_2$, we may form

$$\mathbf{J}=\mathbf{J}_1+\mathbf{J}_2.$$

By an extension of the argument above, it can be shown that, for given values of j_1 and j_2, the allowed values of j are

$$j=j_1+j_2, j_1+j_2-1, \ldots, |j_1-j_2|,$$

so that

$$|j_1-j_2|\leqslant j\leqslant j_1+j_2.$$

C.4 The deuteron

The total intrinsic spin $\mathbf{S}$ of two spin $\frac{1}{2}$ fermions is

$$\mathbf{S}=\mathbf{s}_1+\mathbf{s}_2,$$

where from the rules above the quantum number S can take the values $S=1$ and $S=0$. Explicitly, the three $S=1$ states $|S, S_m\rangle$ are found to be

$$\begin{aligned} |1,1\rangle &= |+\tfrac{1}{2}\rangle_1|+\tfrac{1}{2}\rangle_2 \\ |1,0\rangle &= (|+\tfrac{1}{2}\rangle_1|-\tfrac{1}{2}\rangle_2+|-\tfrac{1}{2}\rangle_1|+\tfrac{1}{2}\rangle_2)/\sqrt{2} \\ |1,-1\rangle &= |-\tfrac{1}{2}\rangle_1|-\tfrac{1}{2}\rangle_2 \end{aligned} \tag{C.6}$$

and the $S=0$ state is

$$|0,0\rangle=(|+\tfrac{1}{2}\rangle_1|-\tfrac{1}{2}\rangle_2-|-\tfrac{1}{2}\rangle_1|+\tfrac{1}{2}\rangle_2)/\sqrt{2}. \tag{C.7}$$

(The factors $\sqrt{2}$ are for normalisation.)

The deuteron is a neutron–proton bound pair having total spin $\mathbf{J}=\mathbf{L}+\mathbf{S}$ with quantum number $j=1$ and total intrinsic spin $\mathbf{S}$ with quantum number $S=1$. Neglecting a small $l=2$ wave component, its spatial wave-function is an $l=0$ state. The wave-function of a deuteron at rest is therefore approximately of the form

$$u(r)|1, m_s\rangle,$$

where r is the distance between the two nuclei.

From (C.6), it will be seen that this wave-function is symmetric under the interchange of proton and neutron. Thus such a state is not accessible to two protons, or to two neutrons, since the wave-functions of two identical fermions must be anti-symmetric under particle interchange (§ 1.1). Although two nucleons with net intrinsic spin zero experience a strong attraction, this attraction is not sufficient to produce a bound state and the deuteron is the only bound state of two nucleons.

Problems

C.1 Show that $l=0$ wave-functions $\psi(r)$ (functions only of the radial coordinate r) are also eigenstates of L_x, L_y, L_z.

C.2 Explain why the single particle states specified by (l, s, j, j_z) introduced in §C.3 have parity $(-1)^l$.

C.3(*a*) Show that the state $|0, 0\rangle$ given by equation (C.7) is an eigenstate of $S_x(=s_{1x}+s_{2x})$, S_y and S_z and hence that it has total spin zero.

(*b*) Show that

$$S_z|1, 1\rangle = \hbar|1, 1\rangle$$

and

$$S^2|1, 1\rangle = 2\hbar^2|1, 1\rangle.$$

Appendix D
Unstable states and resonances

In discussing unstable states, we have in mind a system like an excited nucleus, or a β-unstable nucleus. An unstable state of a system will decay, and often there are several alternative modes of decay. For example, an excited state of a nucleus can have several states of lower energy to which it can decay by emitting a photon, and some β-unstable nuclei can decay by either β^+ or β^--emission. Such distinct modes of decay are called *decay channels*.

An unstable state has certain probabilities per unit time, called partial decay rates, to decay into any of its channels. We shall denote these probabilities by $1/\tau_i$, where the τ_i have the dimension of time and i labels the ith decay channel. The total decay rate $1/\tau$ is the sum of the partial decay rates:

$$\frac{1}{\tau}=\sum_i \frac{1}{\tau_i}. \tag{D.1}$$

We shall also find it useful to define *partial widths* Γ_i and *total widths* Γ by $\Gamma_i=\hbar/\tau_i$, $\Gamma=\hbar/\tau$. These have the dimensions of energy, and clearly

$$\Gamma=\sum_i \Gamma_i. \tag{D.2}$$

The probability that the unstable state will decay to the ith channel is the ratio of the partial decay rate into that channel to the total rate, i.e. Γ_i/Γ.

For many of our applications it will be important that Γ is a small energy on the nuclear energy scale of MeV. For example, in γ-decay a mean life $\sim 10^{-14}$ s corresponds to $\Gamma \sim 0.1$ eV.

We have seen (§2.3) that a decay rate $1/\tau$ implies that a state will decay according to the exponential law

$$P(t)=P(0)e^{-t/\tau},$$

where $P(t)$ is the probability of the state surviving at time t. Thus we can identify the total decay rate with the inverse of the mean life.

D.1 Time development of a quantum system

We denote the wave-function of the unstable state by ψ_0, and the states into which it can decay by $\psi_1, \psi_2, \ldots, \psi_m, \ldots$. (For example, the state ψ_0 might be that of a nucleus prior to α-decay, and the states ψ_m $(m>0)$ describe the residual nucleus and α-particle in their ground states, and the energy and direction of their relative motion.) We shall use 'box normalisation', supposing our system enclosed in a large volume V, so that all the states are discrete and may be normalised to unity. They can always be chosen to be orthogonal to each other. We may therefore take

$$\int \psi_m^* \psi_n \, dq = \delta_{mn},$$

where $dq \equiv$ d(all relevant coordinates).

ψ_0 is not an exact energy eigenstate: if it were, it would not decay. Thus the state $\Psi(t)$ of the system, which is ψ_0 at $t=0$, develops an admixture of the final states. We can express $\Psi(t)$ as a superposition of the states ψ_m, and write

$$\Psi(t)=\sum_{m=0}^{\infty} a_m(t)e^{-iE_m t/\hbar}\psi_m. \tag{D.3}$$

The phase factors, with

$$E_m=H_{mm}=\int \psi_m^* H\psi_m \,\mathrm{d}q,$$

where H is the Hamiltonian of the system, have been inserted for convenience. If all the states were exact eigenstates of H, the coefficients a_m would not depend on time. However, we are interested in the case when the matrix elements $H_{mn}=\int \psi_m^* H\psi_n \,\mathrm{d}q$ are in general non-vanishing for $m\neq n$. Inserting the expansion (D.3) in the Schrödinger equation

$$i\hbar\frac{\partial\Psi}{\partial t}=H\Psi$$

gives

$$\sum_m (i\hbar\dot{a}_m e^{-iE_m t/\hbar}\psi_m+E_m a_m e^{-iE_m t/\hbar}\psi_m)=\sum_m a_m e^{-iE_m t/\hbar}H\psi_m.$$

Multiplying by ψ_n^* and integrating, the orthogonality relation picks out the time dependence of a_n:

$$i\hbar\dot{a}_n=\sum_{m\neq n} H_{nm}e^{-i(E_m-E_n)t/\hbar}a_m \tag{D.4}$$

(noting $E_n=H_{nn}$).

So far our equations are exact. The initial conditions at $t=0$ are $a_0(0)=1$, $a_m(0)=0$ for $m\geqslant 1$.

We now work to first order in the quantities H_{nm}, supposed small when $n\neq m$. Then for $n\geqslant 1$ we have approximately

$$i\hbar\dot{a}_n=H_{n0}(e^{-i(E_0-E_n)t/\hbar})a_0 \tag{D.5}$$

The state ψ_0 is unstable. We make the *ansatz* that $a_0(t)=e^{-\Gamma t/2\hbar}$ so that $|a_0(t)|^2=e^{-t/\tau}$, and the probability of finding the system in the state ψ_0 decays exponentially with time. The equations (D.5) can then be integrated to give

$$\begin{aligned} i\hbar a_n(t)&=H_{n0}\int_0^t e^{-i[(E_0-E_n)-i\Gamma/2]t'/\hbar}\,\mathrm{d}t' \\ &=\frac{\hbar}{i}H_{n0}\left\{\frac{e^{-i[(E_0-E_n)-i\Gamma/2]t/\hbar}-1}{(E_n-E_0)+i\Gamma/2}\right\}. \end{aligned}$$

For times $t\gg\hbar/\Gamma$, $e^{-\Gamma t/2\hbar}\to 0$ and for such times

$$a_n(t)=\frac{H_{n0}}{(E_n-E_0)+i\Gamma/2}.$$

Thus the probability of decay to the state ψ_n is

$$|a_n(t)|^2 = \frac{2\pi}{\Gamma}|H_{n0}|^2 P(E_n - E_0),$$

where

$$P(E_n - E_0) = \frac{\Gamma}{2\pi}\frac{1}{(E_n - E_0)^2 + \Gamma^2/4}.$$

The function $P(E - E_0)$ is shown graphically in Fig. D1. The factor $\Gamma/2\pi$ has been inserted so that

$$\int_{-\infty}^{\infty} P(E - E_0)\,\mathrm{d}E = 1.$$

An important aspect of our result which is exhibited in this figure is that the energy of the final state E_n is not identically equal to E_0, and indeed is not absolutely determined. This feature is not to be interpreted as a violation of energy conservation, but as a consequence of the fact that the state ψ_0 does not have a definite energy. The instability of the state implies that it has a small spread of energy of width Γ about its mean energy $E_0 = \int \psi_0^* H \psi_0\,\mathrm{d}q$. The function $P(E - E_0)$ can be regarded as the probability distribution in energy of the state ψ_0.

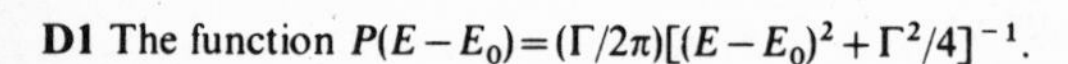

D1 The function $P(E - E_0) = (\Gamma/2\pi)[(E - E_0)^2 + \Gamma^2/4]^{-1}$.

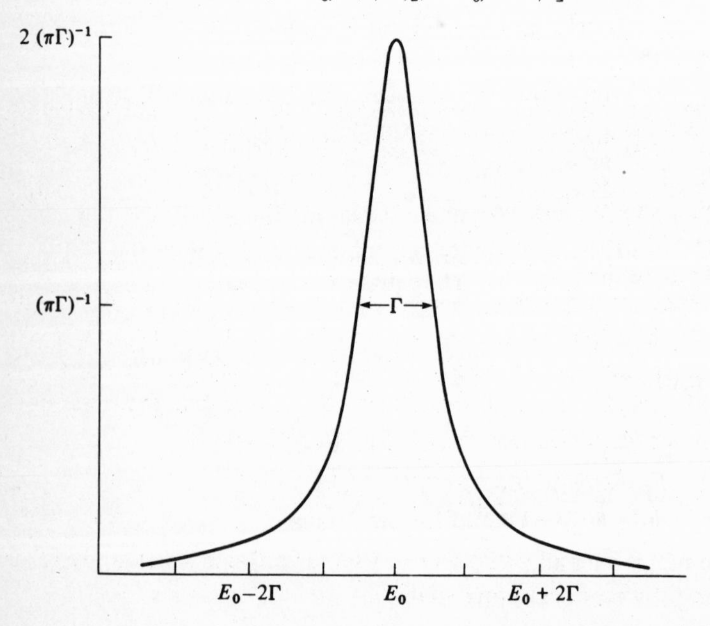

It is interesting to remark that we can now interpret the relationship

$$\tau\Gamma = \hbar$$

as a relationship between uncertainty in energy and lifetime, somewhat similar to the Heisenberg uncertainty relation between momentum and position.

To obtain the probability of decay to a channel i, we must sum over all the states n in i. For example, consider the α-decay of ^{238}U nucleus at rest:

$$^{238}_{92}\mathrm{U} \to {}^{234}_{90}\mathrm{Th} + {}^{4}_{2}\mathrm{He}.$$

In this case, i is the α-decay channel. All the nuclei involved have zero spin, so that the states n in channel i are completely specified by their energy, and the direction of emission of the α-particle. Since there are no spin orientations to be considered, the matrix element H_{n0} will not depend on the direction of emission, and the probability of decay to channel i is

$$\sum_{n \text{ in } i} |a_n(t)|^2 = \frac{2\pi}{\Gamma} \int |H_{n0}|^2 P(E - E_0) n_i(E)\, \mathrm{d}E,$$

where $n_i(E)$ is the density of states in channel i at energy E (Appendix B).

For Γ small, the integral comes almost entirely from around the peak in $P(E - E_0)$ at E_0. Assuming that $n_i(E)$ and $|H_{n0}|^2$ vary slowly with E over the width of the peak, we may evaluate them at E_0 and treat them as constant in the integration to give

$$\sum_{n \text{ in } i} |a_n(t)|^2 = \frac{2\pi}{\Gamma} |H_{n0}|^2 n_i(E_0).$$

Since the probability of decay to channel i is simply Γ_i/Γ, it follows that the partial decay rate, when no spins are involved, is

$$\frac{1}{\tau_i} = \frac{\Gamma_i}{\hbar} = \frac{2\pi}{\hbar} |H_{n0}|^2 n_i(E_0). \tag{D.6}$$

In the more general situation when the decay products have spin, and the initial unstable state has spin j, we will include in the channel i all the spin states of the final particles, and consider the case when the spin of the unstable state is not polarised in any particular direction. We must then average over all $(2j+1)$ initial spin states. After averaging, the result does not depend on direction and the formula (D.6) becomes

$$\frac{1}{\tau_i} = \frac{\Gamma_i}{\hbar} = \frac{2\pi}{\hbar} \frac{n_i(E_0)}{2j+1} \sum |H_{n0}|^2, \tag{D.7}$$

where the sum is over all initial spin states and final spin states, and $n_i(E)$ is the density of states neglecting spin.

D.2 The formation of excited states in scattering: resonances and the Breit–Wigner formula

We now consider the case of a particle i (for example, a neutron) interacting with a nucleus I at an energy close to an energy at which the two can combine to form the unstable excited state X^*. If created, X^* will then decay into one of its decay channels, say channel f. The overall process can be represented by

$$i+I \to X^* \to (\text{channel } f).$$

Such scattering processes which proceed through an intermediate unstable state have important characteristics we wish to discuss. We consider a situation where initially the amplitude a_0 of the unstable state is zero, i.e. $a_0(0)=0$, and the system is in an initial state, ψ_1 say, which belongs to channel i, so that $a_1(0)=1$. The amplitude a_0 develops in time according to the exact equation (D.4) with $n=0$:

$$i\hbar\dot{a}_0=\sum_{m\neq 0} H_{0m}e^{-i(E_m-E_0)t/\hbar}a_m.$$

Again working to first order in the small quantities H_{0m} we have

$$i\hbar\dot{a}_0=-i(\Gamma/2)a_0+H_{01}e^{-i(E_1-E_0)t/\hbar}.$$

The term involving Γ which we have introduced gives the decay of the unstable state in accordance with our *ansatz* and takes account, in a phenomenological way, of the small terms in the exact equation that have otherwise been neglected.

We can write this equation as

$$i\hbar\frac{\mathrm{d}}{\mathrm{d}t}(a_0e^{\Gamma t/2\hbar})=H_{01}e^{-i(E_1-E_0+i\Gamma/2)t/\hbar},$$

so that

$$i\hbar a_0e^{\Gamma t/2\hbar}=\int_0^t H_{01}e^{-i(E_1-E_0+i\Gamma/2)t'/\hbar}\,\mathrm{d}t'.$$

For times t long compared with $\hbar/\Gamma$ we obtain

$$a_0(t)=\frac{H_{01}e^{-i(E_1-E_0)t/\hbar}}{E_1-E_0+i\Gamma/2},$$

and the probability of finding the state ψ_0 is

$$|a_0(t)|^2=\frac{|H_{01}|^2}{(E_1-E_0)^2+\Gamma^2/4}.$$

The decay rate into the channel f is therefore

$$|a_0(t)|^2\frac{1}{\tau_f}=\frac{|H_{01}|^2}{(E_1-E_0)^2+\Gamma^2/4}\left(\frac{\Gamma_f}{\hbar}\right). \tag{D.8}$$

Suppose, for the moment, that the initial particles and the excited state have spin zero. Then it is useful to define

$$\Gamma_i(E) = 2\pi |H_{10}|^2 n_i(E) \tag{D.9}$$

which is a generalisation of (D.6), and, since $|H_{10}|^2 = |H_{01}|^2$, we can re-write the decay rate into channel f as

$$\frac{1}{2\pi\hbar}\frac{1}{n_i(E_1)}\frac{\Gamma_i(E_1)\Gamma_f}{(E_1 - E_0)^2 + \Gamma^2/4}.$$

If the relative motion of the interacting particles is given by the wave-function $V^{-\frac{1}{2}}e^{i\mathbf{k}\cdot\mathbf{r}}$ (in the centre-of-mass coordinate system; see Appendix A), the flux of particles is given by (particle density) × velocity $= V^{-1}v$. The cross-section $\sigma(1 \to f)$ for scattering into channel f is defined by

$$(\text{flux of } 1) \times \sigma(1 \to f) = \text{decay rate into channel } f.$$

Hence

$$\sigma(1 \to f) = \frac{V}{v}\frac{1}{2\pi\hbar}\frac{1}{n_i(E_1)}\frac{\Gamma_i\Gamma_f}{(E_1 - E_0)^2 + \Gamma^2/4}.$$

The density of states in the ith channel is given by (Appendix B)

$$n_i(E)\,\mathrm{d}E = \frac{V}{(2\pi)^3}\,4\pi k^2\,\frac{\mathrm{d}k}{\mathrm{d}E}\,\mathrm{d}E,$$

and $\mathrm{d}E/\mathrm{d}k = \hbar^2 k/m = \hbar v$. ($m$ is the reduced mass of the particles.) Hence, substituting, we obtain

$$\sigma(1 \to f) = \frac{\pi}{k_1^2}\frac{\Gamma_i\Gamma_f}{(E_1 - E_0)^2 + \Gamma^2/4}. \tag{D.10}$$

This result is the *Breit–Wigner formula* for the special case when the incoming particles and the compound nucleus have zero spin. For small Γ, the cross-section peaks sharply at $E_1 = E_0$. The phenomenon is known as resonance scattering and is common in nuclear physics; experimental resonance peaks can often be well fitted by an expression of this form.

The formula for the general spin case is more complicated. Suppose the initial spins of the particles are s_1 and s_2. For example, for neutrons interacting with ^{235}U the spin of the neutron is $s_1 = \frac{1}{2}$ and the spin of ^{235}U is $s_2 = \frac{7}{2}$. If, as in a nuclear reactor, the neutrons and the uranium nuclei are not polarised, then we have to average the cross-section over the $(2s_1 + 1) \times (2s_2 + 1)$ initial spin states. Consider also the formation of an excited state of ^{236}U with spin j. Any of its $(2j + 1)$ sub-states can be formed, and they all contribute to the production of the final state. Equation (D.8) which gives the decay rate into channel f (a fission channel, for example) has to be modified to:

Decay rate into channel f

$$=\frac{1}{(2s_1+1)(2s_2+1)}\frac{\Gamma_f}{\hbar}\frac{1}{(E_1-E_0)^2+\Gamma^2/4}\sum_{\text{spins}}|H_{01}|^2.$$

This, using (D.7) and (D.9), yields the general Breit–Wigner formula:

$$\sigma(1\to f)=\frac{\pi}{k_1^2}\frac{(2j+1)}{(2s_1+1)(2s_2+1)}\frac{\Gamma_i\Gamma_f}{(E_1-E_0)^2+\Gamma^2/4}. \tag{D.11}$$

The total cross-section is obtained by summing over all channels f:

$$\sigma_{\text{tot}}=\frac{\pi}{k_1^2}\frac{(2j+1)}{(2s_1+1)(2s_2+1)}\cdot\frac{\Gamma_i\Gamma}{(E_1-E_0)^2+\Gamma^2/4}. \tag{D.12}$$

The mechanism of formation of the unstable states, and their subsequent decay, is discussed in more detail in Chapter 7.

This expression is a good approximation when one unstable state dominates the cross-section, but scattering which proceeds by other mechanisms than the formation of an unstable state is not included in our discussion. For example, direct reactions, mentioned in Chapter 7, are not included.

Problems

D.1 The Breit–Wigner formula of §D.2 was derived for a particle incident upon a nucleus. It has to be modified if, as in the case of α–α scattering, the 'particle' and the nucleus are identical. Show that for α–α resonant scattering through the formation of ${}^8_4\text{Be}$

$$\sigma=\frac{2\pi}{k_1^2}\frac{\Gamma^2}{(E_1-E_0)^2+\Gamma^2/4}.$$

(See Problem B.1.)

Appendix E
Radioactivity and radiological units

E.1 Becquerels (and curies)

Radioactive nuclei may emit α-particles, electrons, positrons, photons, or fission products. The *activity* of a given nuclear species in a given sample is the average number of decays per second of that species, and is measured in *becquerels*: 1 Bq corresponds to an average of one decay per second.

The total activity of a newly prepared sample may initially increase with time, since the daughter products of a radioactive nucleus may also be radioactive (Problem E.2), though ultimately the total activity must decay to zero.

The becquerel is the SI unit which has replaced the *curie*: $1\,\text{Ci} = 3.7 \times 10^{10}\,\text{Bq}$. The curie was defined originally as the ^{226}Ra activity of a source containing 1 g of ^{226}Ra. Since the mean life of ^{226}Ra is 7.28×10^{10} s, it is easy to check that the definition above is approximately consistent with the older definition.

E.2 Grays and sieverts (and rads and rems)

The absolute *absorbed dose of radiation* at any point in a material is defined as the energy per unit volume that has been absorbed by the material, divided by the mass density at the point. The SI unit for absorbed dose is the *gray*, corresponding to one joule of absorbed energy per kilogram of material. An older unit is the *rad*: $1\,\text{Gy} = 10^2$ rad. In practice, a quoted absorbed dose will be an average over some region, for example a whole body average or an average over some particular organ of the body.

It has been found that radiation damage to living tissue is not simply proportional to the absolute absorbed dose, but depends on several other factors, of which one is the type of radiation. For example, for the same number of grays, α-particles are more damaging than γ-radiation. From medical experience, different types of radiation have been given *relative biological effectiveness factors*. These RBE factors are dimensionless numbers. For many purposes it is conventional to take these factors to be

1 for X-rays, γ-rays and β-particles,
10 for neutrons and protons,
20 for α-particles.

More precisely, the RBE factors depend upon the particle energy as well as on particle type.

The *sievert* is a unit combining the RBE factor with the absolute absorbed dose: the dose equivalent in Sv equals the dose in Gy, multiplied by the appropriate RBE factor for the radiation involved. The dose in Sv is an indicator of the potential harm to living tissue of a given dose of radiation. In practice doses are usually quoted in millisieverts.

The rem is related to the rad in the same way that the sievert is related to the gray, so that $1\,\mathrm{Sv} = 10^2$ rem.

E.3 Levels of radiation

Life has always been subject to the hazards of ionising radiation. There are three principal natural sources: cosmic rays, radioactive nuclei which participate in the chemistry of the body, and radioactive elements present in rocks and soil.

Cosmic rays are very high energy particles which permeate the galaxy. Those which strike the earth's atmosphere cause showers of secondary particles; at sea level these secondaries deliver a dose of about 0.25 mSv per year to the human body (Problem E.3). The precise dose depends on latitude and increases with altitude. At a height of 4000 m the dose would be about 2 mSv per year.

The most significant radioactive nucleus that is found in the body is ^{40}K. Potassium accounts for about 0.2 % of total body weight. The isotope $^{40}_{19}$K, which has spin and parity 4^-, has a long mean life of 1.8×10^9 years, and that which remains since the earth's formation constitutes 0.0117 % of natural potassium. It is an odd–odd nucleus and can undergo all three types of β-decay, but the most common mode (89 %) is electron emission with a kinetic energy release of 1.32 MeV; the remaining 11 % of decays are mostly by electron capture to an excited state of ^{40}Ar, which then itself decays by emitting a 1.46 MeV γ-ray. From these decays the body receives a dose of 0.17 mSv per year. Other radioactive nuclei in the body give in total a contribution of similar magnitude.

The dose of γ-radiation arising from the decay products of radioactive elements in the ground, principally from uranium and thorium, depends on the local geology and is far more variable. Typically the γ-radiation dose is between 0.2 mSv and 0.4 mSv per year, but in areas of granite rock may be several times higher. A greater hazard can arise from the inhalation of the isotopes ^{222}Rn and ^{220}Rn of the inert gas radon. These are decay products of uranium and thorium and being gaseous can diffuse out into the air. In particular they may emanate from some building materials, and accumulate in ill-ventilated rooms. ^{222}Rn decays to a sequence of α-emitters (Table 6.1) which are solids and remain deposited in the lungs. ^{220}Rn, arising from the ^{232}Th chain, is similarly damaging. The dose received depends on building materials and construction, subsoil, and ventilation, and obviously varies widely; it has been estimated that the equivalent whole body dose averages about 1.0 mSv per year.

The natural background radiation thus totals around 2 mSv per year. In the twentieth century man's activities have added somewhat to this. In the UK, the medical uses of radioactivity and X-rays have been estimated as equivalent to 0.5 mSv per year on average (with very wide variations). The average dose due to the radioactive fallout from nuclear weapons testing in the atmosphere is estimated now to be about 0.01 mSv per year; this is about 1% of the peak 1963 value.

Many individuals, through their work in medicine or in nuclear-related industries, are habitually exposed to higher levels of radiation than the average. It is necessary to monitor and protect these people in so far as knowledge and experience will allow. The US maximum permissible occupational whole body dose is at present set at 50 mSv per year.

The gray and sievert are large units in terms of biological damage: whole body doses of about 5 Gy are likely to result in a death rate of 50%. At very low dose levels it is not yet established with certainty whether or not a threshold for biological damage exists. Risks are usually assessed on the assumption of a proportionality between dose and effects. If this is true, it would suggest that the exposure of a population of a million to a whole body dose of 1 mSv results in about twelve fatal cancers, and a similar number of curable cancers. However, the numbers involved are so small that the rule has not been established.

Problems

E.1 5.9% of all ^{235}U fissions produce a ^{137}Cs nucleus within about 5 minutes. The mean life of ^{137}Cs is 44 years. It is a particularly dangerous radioactive isotope if released in the atmosphere.

Estimate the activity of ^{137}Cs in a reactor that has been running at 3 GW thermal power for one year. In the Chernobyl accident 13% of this isotope was released. Estimate its mean activity per square metre if it was spread over a million square kilometres.

E.2 The mean life of $^{226}_{88}$Ra (2300 years) is so long that the radium activity of a newly prepared one curie source will be essentially constant. The mean life of its daughter nucleus $^{222}_{86}$Rn is $\tau = 5.52$ days. Show that the radon activity approaches the radium activity according to

Rn activity $= (1 - e^{-t/\tau})$ Ci.

The subsequent decays in the chain (see Table 6.1) down to $^{210}_{82}$Pb all have mean lives of less than an hour, but $^{210}_{82}$Pb is relatively stable with a mean life of 30 years. Estimate the total activity of the source one month after preparation.

E.3 At sea level most of the flux of ionising particles induced by cosmic rays consists of muons, and over all angles the total flux is about 170 particles $m^{-2} s^{-1}$. The mean muon energy is about 2×10^3 MeV. In the body they will lose energy predominantly by ionisation. Estimate the annual body dose due to this source, taking the mean L of equation (13.4) to be 14.

E.4 The body contains about 18% by weight of carbon, of which a small proportion is the β-unstable isotope $^{14}_{6}C$. About one-third of the decay energy of 0.156 MeV is taken by the electron, and there are no associated γ-rays. The activity of 1 g of natural carbon is 15.3 decays per minute. Estimate the annual whole body dose of radiation from this source.

E.5 Check that the quoted value of 0.17 mSv per year body dose from ^{40}K is consistent with the information given. (Assume that about half of the 1.32 MeV energy release in e^- emission is taken by the anti-neutrino, and take the attenuation length of a 1.45 MeV γ-ray in the body to be 17 cm.)

Further reading

There are many texts on nuclear physics at a more advanced level than this one, for example:

Fermi, E. (1953), *Nuclear Physics*, Chicago: University of Chicago Press. A classic text, full of insights.

Bowler, M. G. (1973), *Nuclear Physics*, Oxford: Pergamon. Intended for undergraduates and gives arguments in greater detail than most texts.

Enge, H. (1966), *Introduction to Nuclear Physics*, Reading, Mass.: Addison-Wesley. A clearly written graduate text.

The student may also find interesting:

Cameron, I. R. (1982), *Nuclear Fission Reactors*, New York: Plenum.

Clayton, D. D. (1968), *Principles of Stellar Evolution and Nucleosynthesis*, New York: McGraw-Hill.

Pochin, E. (1983), *Nuclear Radiation: Risks and Benefits*, Oxford: Clarendon Press.

Answers to problems

(Unless otherwise stated, the mass of a nucleus of mass number A is approximated as A amu.)

Chapter 1

1.1 Ratio $= Gm_e^2(4\pi\varepsilon_0/e^2) = 2.4 \times 10^{-43}$.

1.2(*b*) (*i*) $+1$, (*ii*) -1, (*iii*) -1.

1.3(*b*) Wavelength $\lambda = 2\pi c/\omega = 2\pi(\hbar c)/(\hbar\omega) = 2\pi(197$ MeV fm)/(1 MeV) = 1240 fm.

Chapter 2

2.1 Group velocity $= d\omega/dk = c^2k/\omega = c^2\hbar k/\hbar\omega = c^2p/E$.
For a particle of velocity v, $E = \gamma mc^2$ and $p = \gamma mv$.
Hence group velocity = particle velocity.

2.2 From equation (2.8) the electrostatic energy is of order of magnitude $e^2/4\pi\varepsilon_0 a_0$.
From equation (2.13) and after, the weak interaction energy is of order of magnitude $\varepsilon_0^{-1}(\hbar/M_Z c)^2 e^2/a_0^3$.
Ratio $= 4\pi(\hbar/a_0 M_Z c)^2 \sim 10^{-15}$.

2.3 By momentum conservation the momenta of the two photons must be equal in magnitude (and opposite in direction). They will therefore have equal energy.

2.4 If an e^+e^- pair is created then there is a frame of reference (the centre of mass frame) in which the total momentum of the pair is zero. The photon would therefore have zero momentum and hence zero energy: energy conservation would be violated.

2.5 Since $E = 1$ MeV is much less than the rest energy of the muon we may use non-relativistic mechanics, and the velocity $v = c\sqrt{(2E/m_\mu c^2)} = 4.1 \times 10^7\ \mathrm{ms^{-1}}$. In time τ_μ the muon travels $v\tau_\mu = 90$ m.

2.6(*a*) Not allowed. Such a process need not violate the conservation laws of energy, momentum, angular momentum or electric charge, but it would violate the conservation laws of electron number and muon number. Although searched for, this decay has never been seen. (*b*) and (*c*) can occur.

2.8 Consider a point charge e at position $\mathbf{R}$; then

$$\mathbf{E}(\mathbf{r}) = \frac{e}{4\pi\varepsilon_0}\frac{\mathbf{r}-\mathbf{R}}{|\mathbf{r}-\mathbf{R}|^3}.$$

Under reflection in the origin, $\mathbf{r} \to \mathbf{r}' = -\mathbf{r}$, $\mathbf{R} \to \mathbf{R}' = -\mathbf{R}$ and

$$\mathbf{E}'(\mathbf{r}') = \frac{e}{4\pi\varepsilon_0}\frac{\mathbf{r}'-\mathbf{R}'}{|\mathbf{r}'-\mathbf{R}'|^3} = -\mathbf{E}(\mathbf{r}).$$

The magnetic field due to a current I in a loop is

$$\mathbf{B}(\mathbf{r}) = \frac{\mu_0 I}{4\pi}\int\frac{\mathrm{d}\mathbf{R}\times(\mathbf{r}-\mathbf{R})}{|\mathbf{r}-\mathbf{R}|^3}.$$

Under reflection the vector product does not change sign. Hence $\mathbf{B}'(\mathbf{r}') = +\mathbf{B}(\mathbf{r})$; $\mathbf{B}(\mathbf{r})$ is an axial vector field.

Chapter 3

3.1 The nucleon magnetic dipole moments are vectors aligned with the nucleon spin. In the deuteron the spins are parallel and the moments add to give a net magnitude $\mu_\mathrm{p} + \mu_\mathrm{n} = (2.79284 - 1.91304)e\hbar/2m_\mathrm{p} = 0.8798e\hbar/2m_\mathrm{p}$ (from equation (3.2)).
The measured magnitude is

$$\mu_\mathrm{d} = 0.8574e\hbar/2m_\mathrm{p} = (1 - 0.026)(\mu_\mathrm{p} + \mu_\mathrm{n}).$$

The discrepancy (0.026) could be due to a contribution to the magnetic moment from the orbital motion of the nucleons, associated with the small d-wave component of the deuteron wave-function. (Appendix C, §C.4.)

3.2(*a*) The magnetic field at distance $\mathbf{r}$ from dipole (1) is

$$\mathbf{B} = -\left(\frac{\mu_0}{4\pi}\right)\nabla\left(\frac{\mu\boldsymbol{\sigma}_1\cdot\mathbf{r}}{r^3}\right) = \left(\frac{\mu_0}{4\pi}\right)\left(-\frac{\mu\boldsymbol{\sigma}_1}{r^3} + \frac{3(\mu\boldsymbol{\sigma}_1\cdot\mathbf{r})\mathbf{r}}{r^5}\right).$$

The energy of dipole (2) at $\mathbf{r}$ in this field is

$$-\mu\boldsymbol{\sigma}_2\cdot\mathbf{B}=-\left(\frac{\mu_0}{4\pi}\right)\frac{\mu^2}{r^3}\Omega_{\mathrm{T}}.$$

3.3 Subtracting 1 MeV from the rest energies of the charged particles gives

(udd) 940, (uud) 937; 3 MeV for the extra d quark;

(dds) 1196, (uds) 1192, (uus) 1188; 4 MeV for the extra d quark;

(d$\bar{\mathrm{s}}$) 498, (u$\bar{\mathrm{s}}$) 493; 5 MeV for the extra d quark.

Interchanging a d quark for a u quark always increases the rest energy, in this sample by an average of 4 MeV.

3.4(*a*) Not allowed. Does not conserve electric charge.

(*b*) Not allowed. Does not conserve baryon number or electron number.

(*c*) Not allowed. Does not conserve baryon number.

(*d*) Allowed.

3.5

$\Xi^{*-}\to K^-+\Sigma^0$	Strong. Does not require the weak or the electromagnetic interaction.
$\Sigma^0\to\Lambda+\gamma$	Electromagnetic.
$\Lambda^0\to p+e^-+\bar{\nu}_e$	Involves an anti-neutrino, therefore weak.
$K^-\to\pi^-+\pi^0$	An s quark changes to a d quark, therefore weak.
$\pi^0\to\gamma+\gamma$	Electromagnetic (Fig. 3.6).
$\pi^-\to\mu^-+\bar{\nu}_\mu$	Weak (cf. Fig. 3.5).
$\mu^-\to e^-+\bar{\nu}_e+\nu_\mu$	Weak (Fig. 2.2).

Chapter 4

4.2(*a*) Since $\mathbf{q}=\mathbf{k}_f-\mathbf{k}_i$, where $\mathbf{k}_i$ and $\mathbf{k}_f$ are the initial and final wave vectors, $q^2=k_f^2+k_i^2-2k_fk_i\cos\theta$.

Neglecting the electron rest mass, $k_i=p_i/\hbar=E/\hbar c$, and in scattering from a fixed target there is no energy loss. Hence $k_f=k_i=E/\hbar c$ and $q^2=2E^2(1-\cos\theta)/\hbar^2c^2$, $q=(2E/\hbar c)\sin(\theta/2)$.

4.3 By the uncertainty principle, the mean magnitude of the lepton momentum $p\sim\hbar/a$. But $p=mv$, so that $(v/c)\sim\hbar/amc=e^2/4\pi\varepsilon_0\hbar c=\frac{1}{137}$. The characteristic time $t=a/v=137a/c=1.2\times10^{-19}$ s, and $(\tau_\mu/t)\sim 2\times10^{13}$.

4.4(*b*) This follows from perturbation theory in quantum mechanics. Since the integral is over nuclear dimensions $r\leqslant R$, it is reasonable to approximate $\psi(r)$ by $\psi(0)=\pi^{-\frac{1}{2}}(Z/a)^{\frac{3}{2}}$, which with Problem 4.4(*a*) gives the result.

4.5 The total binding energy of two α-particles is 56.60 MeV, 0.1 MeV greater than the binding energy of $^8_4$Be. $^8_4$Be decays to two α-particles and to conserve energy the 0.1 MeV of nuclear energy is converted into the kinetic energy of their motion. $^{12}_6$C is more strongly bound than three α-particles by 7.26 MeV. The binding energy of $^6_3$Li is 31.99 MeV, 1.47 MeV greater than the total binding energy of $^2_1$H and $^4_2$He. Energy is conserved overall, and the nuclear energy released is taken by the γ-ray and the kinetic energy of the $^6_3$Li.

4.6 Treating A as a continuous variable, the maximum is where $d(B/A)/dA=0$, i.e. at

$$\frac{A}{2}=Z=\frac{b}{d}=25.7.$$

The nearest integer is $Z=26$, which gives the maximum of (B/A).

4.7 For $A=100$, the formula gives $Z=43$. $^{100}_{43}$Tc is an odd–odd nucleus and unstable. Both $^{100}_{42}$Mo and $^{100}_{44}$Ru are stable. For $A=200$, the formula gives $Z=80$, and $^{200}_{80}$Hg is stable.

4.8 Suppose the sample contains N ^{14}C nuclei. Then the mean number of decays per second is $N/\tau=15.3/60\ \text{s}^{-1}$, and hence $N=6.7\times10^{10}$. The atomic mass of natural carbon is 12.01 amu $=2\times10^{-23}$ g. Therefore 1 g of carbon contains 5×10^{22} atoms, and the proportion of ^{14}C in the sample is 1.3×10^{-12}.

4.9 As a rough rule, for A odd there is only one β-stable nucleus and for A even, two. Up to and including $A=209$ there are 105 odd-A nuclei and 104 even-A nuclei, and so about 310 β-stable nuclei. All these have $Z\leqslant 83$ which implies an average of about 3.7 stable isotopes per element.

Chapter 5

5.2(*a*) From equations (5.4) and (5.5), for neutrons with kinetic energy E, $\mathcal{N}(E)=N(E/E_n^F)^{\frac{3}{2}}$.

The density of states is $n(E)=d\mathcal{N}/dE=\frac{3}{2}NE^{\frac{1}{2}}(E_n^F)^{-\frac{3}{2}}$.
Hence the total neutron kinetic energy is

$$\int_0^{E_n^F} En(E)\,dE=\tfrac{3}{5}NE_n^F.$$

A similar result holds for the protons, noting that proton kinetic energies are given by $(E_p-\bar{U})$.

(*b*) Again consider the neutrons; since E_n^F is proportional to $N^{\frac{2}{3}}$ we can write

$$E_n^F=E_0^F(N/N_0)^{\frac{2}{3}}=E_0^F(1+\Delta N/N_0)^{\frac{2}{3}}.$$

The result follows on expansion.

Similarly $E_p^F - \bar{U} = E_0^F(1-\Delta N/N_0)^{\frac{2}{3}}$.

(*c*) $\frac{3}{5}NE_n^F = \frac{3}{5}N_0E_0^F(1+\Delta N/N_0)^{\frac{5}{3}}$

$$\approx \frac{3}{5}N_0E_0^F\left(1+\frac{5}{3}\left(\frac{\Delta N}{N_0}\right)+\frac{5}{9}\left(\frac{\Delta N}{N_0}\right)^2\right).$$

Adding the proton kinetic energy, one obtains the total kinetic energy

$$\tfrac{3}{5}E_0^F A + \tfrac{1}{3}E_0^F\frac{(N-Z)^2}{A}.$$

5.3 $$\bar{U}_c = \frac{(Z-1)e^2}{4\pi\varepsilon_0 R}\left(\frac{4\pi R^3}{3}\right)^{-1}\int_0^R 4\pi r^2\left(\frac{3}{2}-\frac{r^2}{2R^2}\right)\mathrm{d}r$$

$$= \frac{6}{5}\frac{(Z-1)e^2}{4\pi\varepsilon_0 R} \approx 21.5\ \text{MeV} \quad \text{for } ^{208}_{82}\text{Pb}.$$

5.4 Since $Z=N$, the energy due to the strong nucleon–nucleon interaction should be the same for a neutron as for a proton in a similar state; thus $\bar{U}$ should be given by the Coulomb contribution, $\bar{U}_c$ of Problem 5.3(*a*). This yields $\bar{U}=8.7$ MeV. The separation energy is the binding energy at the Fermi level. Assuming $E_p^F \approx E_n^F + \bar{U}$, it follows that $S_p = (15.6-8.7)$ MeV.

5.5 In the simple shell model, $^{31}_{15}$P has an odd proton; Table 5.1 suggests this is in the $2s_{\frac{1}{2}}$ shell, giving nuclear spin and parity $j^P = \frac{1}{2}^+$.

$^{67}_{30}$Zn has an odd neutron. Suggested shell $f_{\frac{5}{2}}$; $j^P = \frac{5}{2}^-$.

$^{115}_{49}$In has an odd proton. Suggested shell $1g_{\frac{9}{2}}$; $j^P = \frac{9}{2}^+$.

These suggestions are all in agreement with experiment.

5.6 The spins and parities of all but $^{26}_{13}$Al are in accord with pairing and shell filling as in Table 5.1. $^{26}_{13}$Al is odd–odd; the model suggests the odd neutron and proton both to be in $\frac{5}{2}^+$ states. Such a configuration would have the measured parity of $^{26}_{13}$Al; and the measured spin of 5 suggests that the spins are paired parallel.

For the magnetic moments, equation (5.26) gives:

$^{43}_{20}$Ca, odd neutron, $l=3$, $j=l+\frac{1}{2}$; $\mu=-1.92\ \mu_N$

$^{93}_{41}$Nb, odd proton, $l=4$, $j=l+\frac{1}{2}$; $\mu=6.80\ \mu_N$

$^{137}_{56}$Ba, odd neutron, $l=2$, $j=l-\frac{1}{2}$; $\mu=1.15\ \mu_N$

$^{197}_{79}$Au, odd proton, $l=2$, $j=l-\frac{1}{2}$; $\mu=0.12\ \mu_N$.

$^{26}_{13}$Al. The fact that the two angular momenta appear to be aligned suggests that we can simply add the Schmidt values to obtain the estimate $\mu=2.9\ \mu_N$.

5.7 For a proton, $j=\frac{1}{2}$, $\mu=2.80\ \mu_N$.

For $^{43}_{20}$Ca, $j=\frac{7}{2}$, $\mu=-1.32\ \mu_N$.

Taking the nuclear magneton $\mu_N = 3.15\times10^{-14}$ MeV T^{-1} and using equation (5.21) gives

$\nu=\omega/2\pi=43$ MHz for protons,

$\nu=\omega/2\pi=2.9$ MHz for $^{43}_{20}$Ca.

5.8 The volume of the ellipsoid is $(4\pi/3)a^2b$ and hence the charge density is $3Ze/4\pi a^2 b$.

$$Q_{zz}=\frac{3Z}{4\pi a^2 b}\iiint(2z^2-x^2-y^2)\,dx\,dy\,dz,$$

where the integral is through the ellipsoid.
Make a change of scale: $x=ax'$, $y=ay'$, $z=bz'$; then

$$Q_{zz}=\frac{3Z}{4\pi}\iiint(2b^2z'^2-a^2x'^2-a^2y'^2)\,dx'\,dy'\,dz',$$

where the integral is now through the unit sphere. Also

$$\iiint x'^2\,dx'\,dy'\,dz'=\frac{1}{3}\int_0^1 r'^2 4\pi r'^2\,dr'=\frac{4\pi}{15},\quad \text{etc.,}$$

giving

$$Q_{zz}=\frac{2Z}{5}(b^2-a^2).$$

Taking the density of nuclear matter to be $\rho_0=0.17$ nucleons fm^{-3} and

$$\frac{4\pi}{3}a^2b\rho_0=A,$$

these equations lead to $b=7.7$ fm, $a=5.6$ fm.

Chapter 6

6.1 $Q=0.094$ MeV, $r_s\approx 2\times 4^{\frac{1}{3}}\times 1.1$ fm $=3.5$ fm,
$r_c=4e^2/4\pi\varepsilon_0 Q=61$ fm.
$r_s/r_c=0.057$ and Fig. 6.3 or equation (6.16) gives $\mathscr{G}=0.70$.
The reduced mass is $m_\alpha/2$. Hence $G=13$.
Taking $\tau_0\approx 7\times 10^{-23}$ s, as in other α-decays, leads to the estimate $\tau=3\times 10^{-17}$ s, though this excellent agreement is fortuitous.

6.2(*a*) To apply equations (6.2) and (6.15) to a positron, $2Z_d\rightarrow Z_d$ and m becomes the positron mass.
Thus r_c = 112 fm. Values of r_s between r_s = 0 and $r_s = 1.1A^{\frac{1}{3}}$ fm = 6.37 fm are reasonable, giving $\mathscr{G}$ = 1, 0.70; G = 1.81, 1.27, and a suppression factor e^{-G} in the range 0.16 to 0.28.

(*b*) $Q=1.8$ MeV, $r_c=123$ fm, $r_s\approx 8.1$ fm, $\mathscr{G}=0.68$, $G=154$,
which with $\tau_0=7\times 10^{-23}$ s gives a partial mean life for α-decay
$\tau_\alpha\sim 10^{37}$ years.
(Age of Solar System $\sim 10^9$ years.)

6.3 In α-decay in material, the kinetic energy is largely converted into heat and N atoms of ^{238}Pu would on average produce $N(5.49\text{ MeV})/\tau$ of power.
For $1\text{ kW} = 6.24 \times 10^{15}\text{ MeV s}^{-1}$ we need $N = 4.6 \times 10^{24}$, or 1.8 kg of ^{238}Pu.
The decay rate of the by-product ^{234}U is so low that the heat from its decay is negligible.
For the remaining mass of plutonium to be 1.8 kg after 50 years requires 2.7 kg initially.

6.4 Suppose that when the sample of rock was formed, say T years ago, it contained no lead but N_1 atoms of ^{238}U (mean life τ_1) and N_2 atoms of ^{235}U (mean life τ_2). Then it would now contain $N_1 e^{-T/\tau_1}$ atoms of ^{238}U and, since each decayed uranium atoms becomes a lead atom, $N_1(1-e^{-T/\tau_1})$ atoms of ^{206}Pb. Setting $(1-e^{-T/\tau_1})/e^{-T/\tau_1} = 0.0797$ suggests $T = 497 \times 10^6$ years. Similarly for ^{235}U and ^{207}Pb,
$(1-e^{-T/\tau_2})/e^{-T/\tau_2} = 0.675$
suggests $T = 531 \times 10^6$ years. (The discrepancy could be due to the effect of water on the rock, for example.)

6.5 Neglecting the excitation energy, the kinetic energy of the fragments can be estimated using equation (6.18), which gives $\Delta B = 178$ MeV. Each fragment would then have velocity $12 \times 10^6\text{ m s}^{-1}$. In the frame in which the fragment is at rest, a 2 MeV neutron has velocity $20 \times 10^6\text{ m s}^{-1}$. In the laboratory frame, the distribution of emitted neutrons is peaked in the direction of the moving fragment.

Chapter 7

7.3(*a*) $E_i + E_0 = E_f +$(excitation energy)+($^{17}O^*$ recoil energy).
Hence the recoil energy is 0.26 MeV and the recoil velocity is $v/c = 5.7 \times 10^{-3}$ (approximating the mass of $^{17}O^*$ by 17 amu).

(*b*) If the photon has energy E_γ it has momentum E_γ/c, and to conserve momentum this must be the recoil momentum of the ^{17}O. Hence the ^{17}O recoil energy E_R is $E_R = (E_\gamma/c)^2/(34\text{ amu})$.
To conserve energy, $E_\gamma + E_R = 0.87$ MeV.
We could solve these equations for E_γ, but clearly E_R is small, and to two significant figures it is sufficient to take $E_\gamma = 0.87$ MeV in the first equation to give $E_R = 24$ eV.

7.3(*c*) The photon energy will be a maximum if it is emitted parallel to the motion of the $^{17}O^*$. By a Lorentz transformation to the laboratory frame

$$E_\gamma^{\text{lab}} = (1 + v/c)E_\gamma/(1 - v^2/c^2)^{\frac{1}{2}}$$

and hence $E_\gamma^{\text{lab}} - E_\gamma = 5$ keV.

Similarly the photon energy will be a minimum if it is emitted anti-parallel, in which case $E_\gamma^{\text{lab}} - E_\gamma = -5$ keV.

7.4 $^{11}_{6}$C is less bound than $^{11}_{5}$B by 2.762 MeV. The difference of Coulomb energies of uniformly charged spheres of net charge $6e$ and $5e$ and radius $R = 1.1 \times 11^{\frac{1}{3}}$ fm is

$$\frac{3}{5}\frac{e^2}{4\pi\varepsilon_0 R}(6^2 - 5^2) = 4 \text{ MeV}.$$

This is a 50% over-estimate of the observed energy difference, and we would need to take $R_c = 1.45\,R$ to obtain agreement. The approximation of a uniform charge distribution is inadequate for precise calculations, especially for light nuclei. In reality some charge is displaced to larger distances (see Fig. 4.3) thereby reducing the energy. Calculations using the more realistic distributions are in better accord with the data.

7.5 The decay by neutron emission with a kinetic energy release of 0.41 MeV need involve only the strong interaction. There is no Coulomb barrier, and only a small angular momentum barrier: to conserve angular momentum and parity the angular momentum of the ^{16}O–n pair must be $l = 1$.
The mean life is still quite long on the nuclear time-scale of $\sim 10^{-22}$ s. $\Gamma = \hbar c/\tau c = 0.04$ MeV.

7.6 The nuclear transition is $\frac{1}{2}^- \to \frac{3}{2}^-$, so the photon will have positive parity and angular momentum quantum number $2 \geqslant j \geqslant 1$. The most likely transition is with $j = 1$, which would be magnetic dipole. The photon energy is about 2.13 MeV. From Fig. 7.6, a rough estimate of the mean life is $\tau \sim 10^{-17} \times (100/A)^{\frac{2}{3}} \times 20A^{\frac{2}{3}}\,\text{s} \sim 4 \times 10^{-15}$ s. (The experimental mean life is 5.2×10^{-15} s.) An electric quadrupole transition with $j = 2$ is also possible, but Fig. 7.6 suggests its partial decay rate to be much slower than the magnetic dipole rate.

7.7 The lowest six energy levels (comprising 26 states) all have positive parity. $^{10}_{5}$B has three protons in the $p_{\frac{3}{2}}$ shell and three neutrons in the $p_{\frac{3}{2}}$ shell. There are many combinations of the single nucleon p-states and they all have positive parity, $(-1)^6$. The lowest observed states can be considered to be constructed from these.
The 1.74 MeV level can decay to the 0.72 MeV level by a magnetic dipole $(0^+ \to 1^+)$ transition with $E_\gamma \approx 1.02$ MeV. This level can in turn decay to ground by an electric quadrupole $(1^+ \to 3^+)$ transition with $E_\gamma \approx 0.72$ MeV. Neglecting internal conversion the ratio of photons emitted is clearly one-to-one.
The 1.74 MeV level can also decay directly to ground with $E_\gamma \approx$ 1.74 MeV but this $(0^+ \to 3^+)$ transition is magnetic octapole and very slow. Using Fig. 7.6 the number of photons emitted with energies 1.02 MeV, 0.72 MeV and 1.74 MeV should be in proportion $1:1:10^{-8}$.

Chapter 8

8.2 A neutron with kinetic energy 0.1 eV has $v/c = 1.46 \times 10^{-5}$ giving $\sigma = 2670$ b, $l \approx 1.56$ cm. The probability of a neutron penetrating a distance x into the gas without interaction is $e^{-x/l}$. For this probability to be 0.1, we require $x = 3.6$ cm. The active region of the detector should be at least of this thickness.

8.3(*a*) $\tau = \hbar/\Gamma = \hbar c/\Gamma c = 1.3 \times 10^{-21}$ s.

(*b*) In this example the elastic width equals the total width to a good approximation, since there is not enough energy to induce other nuclear reactions. The spin of the neutron is $s_1 = \frac{1}{2}$ and the spin of ${}^4_2\mathrm{H}$ is $s_2 = 0$. Hence the statistical factor in the Breit–Wigner formula is $(2j+1)/(2s_1+1)(2s_2+1) = 2$, and the cross-section at energy E is

$$\sigma(E) = \frac{2\pi}{k^2} \frac{\Gamma^2}{(E - E_0)^2 + \Gamma^2/4} \quad \text{(equation (D.11))}.$$

At energy $E = E_0$, $\sigma = 8\pi/k^2 = 4\pi\hbar^2/mE_0$, where m is the reduced mass. Hence $\sigma \approx 3.2$ b.

8.4 The coefficient of the $(1/v)$ term is large if the incident neutron can easily induce a nuclear reaction (as in the case of ${}^{235}\mathrm{U}$ fission), or if there is an excited state close to zero incident neutron energy. Neither of these conditions is apparently satisfied in the case of ${}^{238}\mathrm{U}$. However, one would expect to see a small $(1/v)$ contribution at even lower neutron energies due for example to residual radiative capture.

8.5 The total width Γ is given approximately by $\Gamma = \Gamma_\gamma + \Gamma_n$, and the relative probability of neutron radiative capture is

$\Gamma_\gamma/\Gamma = 1/(1 + \Gamma_n/\Gamma_\gamma)$.

In this application of the Breit–Wigner formula the neutron spin $s_1 = \frac{1}{2}$ and the spin of the even–even nucleus ${}^{238}\mathrm{U}$ is $s_2 = 0$. Also $j = \frac{1}{2}$. Hence at resonance

$$\sigma_\gamma = \frac{4\pi}{k^2} \frac{\Gamma_n \Gamma_\gamma}{(\Gamma_n + \Gamma_\gamma)^2},$$

where $k^2 = 2mE_0/\hbar^2$.

From Fig. 8.5, $E_0 = 6.7$ eV and $\sigma_\gamma = 2 \times 10^4$ b. Hence

$$\frac{\Gamma_n \Gamma_\gamma}{(\Gamma_n + \Gamma_\gamma)^2} = \frac{\Gamma_n/\Gamma_\gamma}{(1 + \Gamma_n/\Gamma_\gamma)^2} = 0.052.$$

Since we are told that Γ_n/Γ_γ is small, we take the solution $\Gamma_n/\Gamma_\gamma = 0.058$. Hence capture is 95% probable.

Chapter 9

9.1 The molecules $CH_4 + 2O_2$ have a mass ≈ 80 amu and release 9 eV in chemical reaction, i.e. 0.11 eV per amu. ${}^{235}\mathrm{U}$ has a mass ≈ 235 amu and releases about 200×10^6 eV on fission (Table 9.1 and discussion), giving 0.85×10^6 eV per amu.

Ratio $\approx 8 \times 10^6$.

9.3 $S_n = B(Z, N) - B(Z, N-1)$. The quoted difference comes from the pairing energy terms. All other terms in the mass formula give contributions to S_n which for a heavy nucleus vary only slowly with A.

9.4 The probability of the neutron inducing fission at the nth collision is $p(1-p)^{n-1}$.
By definition the mean number of collisions is

$$\bar{n} = \sum_1^\infty np(1-p)^{n-1} = -p\frac{\mathrm{d}}{\mathrm{d}p}\sum_1^\infty (1-p)^n.$$

Summing the geometric series

$$\bar{n} = -p\frac{\mathrm{d}}{\mathrm{d}p}[1-(1-p)]^{-1} = \frac{1}{p}.$$

9.5(*a*) If v is the neutron velocity in the laboratory frame, its velocity in the centre-of-mass frame is

$$v - \frac{m_\mathrm{n} v}{M+m_\mathrm{n}} = \frac{Mv}{M+m_\mathrm{n}}.$$

In the centre-of-mass frame it loses no energy on scattering, but suppose it is deflected through an angle θ. In the laboratory frame it will then have a component of velocity $v(m_\mathrm{n} + M\cos\theta)/(M+m_\mathrm{n})$ in the original direction and a perpendicular component $Mv\sin\theta/(M+m_\mathrm{n})$. The result follows on averaging over all angles θ.

(*b*) On average, after N collisions a neutron with initial energy E_0 will have energy $E_N = \alpha^N E_0$, where

$$\alpha = \frac{M^2 + m_\mathrm{n}^2}{(M+m_\mathrm{n})^2} = 0.86.$$

For $E_0 = 2$ MeV, $E_N = 0.1$ eV, the number of collisions required is $N \sim 110$.
The mean time between collisions for a neutron of energy E is $\Delta t = 1/\sigma\rho v = 1/\sigma\rho\sqrt{(2E/m_\mathrm{n})}$, and it loses energy $\Delta E = (1-\alpha)E$.
Approximating the mean rate of change of energy by

$$\frac{\mathrm{d}E}{\mathrm{d}t} \approx -\frac{\Delta E}{\Delta t} = -(1-\alpha)\sigma\rho(2/m_\mathrm{n})^{\frac{1}{2}}E^{\frac{3}{2}}$$

gives the time to 'cool' to $E_N = 0.1$ eV:

$$\text{time} = \frac{1}{(1-\alpha)\sigma\rho c}\sqrt{\left(\frac{m_\mathrm{n}c^2}{2}\right)}\int_{E_N}^{E_0}\frac{\mathrm{d}E}{E^{\frac{3}{2}}} \approx \frac{1}{(1-\alpha)\sigma\rho c}\left(\frac{2m_\mathrm{n}c^2}{E_N}\right)^{\frac{1}{2}} = 8 \times 10^{-5}\ \text{s}.$$

9.6 From the text $l=3$ cm, $l/v=1.5\times10^{-9}$, $\nu=2.5$ and $t_p=9\times10^{-9}$ s. Substituting the form $\rho(r,t)=f(r)e^{\lambda t}$ into the equation yields

$$\lambda f(r)=\frac{(\nu-1)}{t_p}f(r)+\frac{D}{r}\frac{d^2}{dr^2}(rf(r)),$$

which has solutions of the form $f(r)=(1/r)\sin(kr)$ provided

$$\lambda=\frac{(\nu-1)}{t_p}-Dk^2.$$

To satisfy the boundary condition,

$$k(R+0.71l)=n\pi,\quad n=1,2,\ldots$$

To avoid an exponential increase in density $\lambda\leqslant0$ for all n and therefore

$$(R+0.71l)^2\leqslant\frac{t_p}{(\nu-1)}D\pi^2,$$

i.e.

$$R\leqslant\left[\pi\sqrt{\left[\frac{t_pv}{3(\nu-1)l}\right]}-0.71\right]l,$$

and the critical radius in this approximation is 8.8 cm.

9.7(*a*) Suppose that the probability of a neutron induced fission to result in a fragment which produces a delayed neutron is $\delta\nu$, and that the number of such fragments at any time is $N(t)$, then

$$\frac{dN}{dt}=-\frac{N(t)}{\tau_\beta}+\frac{q\,\delta\nu}{t_p}n(t).$$

This equation has the solution

$$N(t)=\frac{q\,\delta\nu}{t_p}\int_{-\infty}^{t}e^{-(t-t')/\tau_\beta}n(t')\,dt'.$$

Including delayed neutrons in equation (9.1) gives

$$\frac{dn}{dt}=\frac{(\nu_q-1)}{t_p}n(t)+\frac{N(t)}{\tau_\beta},$$

which is the equation quoted.

9.7(*b*) $$\lambda=\frac{(\nu q-1)}{t_p}+\frac{\delta\nu q}{t_p(1+\lambda\tau_\beta)}$$

or

$$(\lambda\tau_\beta)^2+(\lambda\tau_\beta)[1-(\nu q-1)\tau_\beta/t_p]-(\tau_\beta/t_p)[(\nu+\delta\nu)q-1]=0.$$

(*c*) Clearly $\lambda=(\nu q-1)/t_p=1$ s when $\delta\nu=0$.

(*d*) Substituting the given values in the quadratic equation gives

$$(\lambda\tau_\beta)^2+781(\lambda\tau_\beta)-10=0.$$

The positive solution for $(\lambda\tau_\beta)$ corresponds to an exponentially increasing $n(t)$, with time-scale $1/\lambda=13$ min.

Chapter 10

10.1(*a*) $6.5\times10^{14}\,\mathrm{m^{-2}\,s^{-1}}$.

(*b*) Neutrino mean free path $l=1/\sigma\rho_{\mathrm{nuc}}$ and $\rho_{\mathrm{nuc}}\leqslant(1\,\text{Å})^{-3}=10^{30}\,\mathrm{m^{-3}}$. Hence $l\sim10^{15}\,\mathrm{km}\sim10^{11}$ Earth diameters.

10.2 Thermal velocities in the gaseous state of hydrogen exceed the escape velocity in the Earth's gravitational field. Only hydrogen that is chemically bound remains.

10.3 4.86 MeV.

10.4(*a*) $\rho_{\mathrm{p}}=3.4\times10^{31}\,\mathrm{m^{-3}},\quad \tau^2e^{-\tau}=2.5\times10^{-4},$

$$\lambda_{\mathrm{pp}}=\overline{v\sigma_{\mathrm{pp}}}=1.4\times10^{-49}\,\mathrm{m^3\,s^{-1}}.$$

The p–p reaction rate $=\frac{1}{2}\lambda_{\mathrm{pp}}\rho_{\mathrm{p}}^2=8.1\times10^{13}\,\mathrm{m^{-3}\,s^{-1}}$.
Each p–p reaction produces 13.1 MeV and hence the contribution to $\varepsilon=170\,\mathrm{W\,m^{-3}}$.

(*b*) The p–^{12}C reaction rate $=\rho_{\mathrm{p}}\rho_{\mathrm{c}}\lambda_{\mathrm{pc}}$, and hence the mean time for one carbon nucleus to react is $1/\rho_{\mathrm{p}}\lambda_{\mathrm{pc}}\approx10^6$ years.

10.5(*a*) The reaction rate per unit volume is $\rho_{\mathrm{d}}^2\overline{\sigma v}$ and each reaction reduces ρ_{d} by one. Since this is the dominant reaction (Fig. 10.4)

$$\frac{\mathrm{d}\rho_{\mathrm{d}}}{\mathrm{d}t}=-\rho_{\mathrm{d}}^2\overline{\sigma v}.$$

(*b*) This equation can be integrated to give

$$\overline{\sigma v}t=\frac{1}{\rho_{\mathrm{d}}}-\frac{1}{\rho_0},$$

and hence the proportion 'burned' is

$$\frac{\rho_0-\rho_{\mathrm{d}}}{\rho_{\mathrm{d}}}=\frac{\rho_0t_{\mathrm{c}}\overline{\sigma v}}{1+\rho_0t_{\mathrm{c}}\overline{\sigma v}}.$$

(*c*) $10^{20}\,\mathrm{m^{-3}\,s}$.

Chapter 11

11.1 All states are occupied up to $k=k_{\mathrm{F}}$ (equation (B.5)). Using equation (B.4) the mean value of k is

$$\bar{k}=\int_0^{k_{\mathrm{F}}}k^3\,\mathrm{d}k\bigg/\int_0^{k_{\mathrm{F}}}k^2\,\mathrm{d}k=(3/4)k_{\mathrm{F}}.$$

For extreme relativistic electrons, $E\approx pc=\hbar ck$; hence the mean energy is $\hbar c\bar{k}=(3/4)\hbar ck_{\mathrm{F}}$.

11.3 The reaction is endothermic and requires an energy input of 0.78 MeV $=Q$. The reaction will proceed if $\varepsilon_{\mathrm{F}}>Q$. (At $T=0$ K the proton Fermi energy is less than the electron Fermi energy $E_{\mathrm{e}}^{\mathrm{F}}$ by a factor $\sim m_e/m_p$.) Using the non-relativistic formula $\varepsilon_F = \hbar^2 k_F^2/2m_e$ for a rough estimate, and equation (B.5), the number density of electrons ρ_e must satisfy

$$\rho_e \geqslant \frac{1}{3\pi^2}\left(\frac{2m_e c^2 Q}{(\hbar c)^2}\right)^{\frac{3}{2}} = 3\times 10^{-9}\,\text{fm}^{-3}.$$

The corresponding hydrogen density is $\sim 5\times 10^9$ kg m^{-3}.

11.4 Taking the Sun's density to be that of nuclear matter, i.e. $\rho_0 = 0.17$ nucleons fm^{-3}, then its mass density would be $\sim 3\times 10^{17}$ kg m^{-3} and radius ~ 12 km.

11.5 If $E > E_0 = 8$ MeV, $k_B T = 0.5$ MeV, then $E/k_B T > 16$. Hence

$$n \approx \frac{1}{\pi^2(\hbar c)^3}\int_{E_0}^{\infty} E^2 e^{-E/k_B T}\,\text{d}E \approx \frac{1}{\pi^2(\hbar c)^3} E_0^2 k_B T e^{-E_0/k_B T},$$

giving $n \approx 5\times 10^{31}$ m^{-3}.

11.6(*a*) From § 10.3,

$$\overline{v\sigma} = \left(\frac{2}{\pi}\right)^{\frac{1}{2}}\left(\frac{m_\alpha}{2k_B T}\right)^{\frac{3}{2}}\int_0^{\infty} v\sigma e^{-E/k_B T} v^2\,\text{d}v$$

where $E = \frac{1}{2}(m_\alpha/2)v^2$, i.e.

$$\overline{v\sigma} = \left(\frac{1}{m_\alpha k_B T}\right)^{\frac{3}{2}} 8\pi^{\frac{1}{2}}\hbar^2\Gamma^2 \int \frac{e^{-E/k_B T}}{(E-E_0)^2 + \Gamma^2/4}\,\text{d}E.$$

Integrating over the narrow resonance peak gives the result.

(*b*) In the plasma, the rate of production of ^{8}Be is $\frac{1}{2}\rho_\alpha^2\overline{v\sigma}$ (equation (10.6)). The rate of decay per unit volume is $\rho_{Be}/\tau = \rho_{Be}\Gamma/\hbar$. In equilibrium these rates are equal.

(*c*) 2.3×10^{-10}.

Chapter 12

12.1 Replacing $S_0(E_e)$ in equation (12.5) by $(E_0 - E_e)^2 E_e^2$, the mean electron energy is clearly $E_0/2$ by symmetry. The mean life is inversely proportional to

$$f(Z_d, E_0) \approx \left(\frac{1}{m_e c^2}\right)^5 \int_0^{E_0} (E_0 - E_e)^2 E_e^2\,\text{d}E_e = \frac{1}{30}\left(\frac{E_0}{m_e c^2}\right)^5.$$

The proportion of decays within ΔE of the end-point is

$$\frac{1}{f(Z_d, E_0)}\left(\frac{1}{m_e c^2}\right)^5 \int_{E_0-\Delta E}^{E_0} (E_0 - E_e)^2 E_e^2\,\text{d}E_e \approx \frac{1}{f}\frac{E_0^2\,\Delta E^3}{3(m_e c^2)^5}.$$

Substituting for *f*, the result follows.

12.2 By momentum conservation

$$p_\nu^2 c^2 = p_{Li}^2 c^2 = 2\times 6536\ \text{MeV}\times(55.9\pm 1.0)\ \text{eV}$$
$$= (0.7307\pm 0.0131)\ \text{MeV}^2,$$

and by energy conservation

$m_\nu^2 c^4 = [(0.862)^2 - 0.7307 \pm 0.0131]$ MeV,

giving $0 \leqslant m_\nu c^2 \leqslant 160$ keV.

12.3 $fT_{\frac{1}{2}} = 4760$ s. For a free neutron $fT_{\frac{1}{2}} = 1015$ s, from § 12.8.
In the simple shell model the 1s neutron and proton spatial wave functions would be the same if Coulomb distortions were neglected, and the spin states similar to those of a free neutron and free proton. Thus the predicted $fT_{\frac{1}{2}}$ value would be the same. However, since $Z = 15$, Coulomb distortions are not insignificant. Also shell model predictions for the Gamow–Teller matrix elements, like those for the similar magnetic moment matrix elements (§ 5.6), are not accurate.

12.4 Note there is no Coulomb factor in the matrix element.
$E_e = 0.71$ MeV; $\sigma = 3.3 \times 10^{-20}$ b.

12.5 ^{11}Be decay: $|\mathbf{R}_{f0}| = 0.7$ fm, a nuclear size.
Atomic decay: $|\mathbf{R}_{f0}| = 0.4$ Å, an atomic size.

Chapter 13

13.1 $$R_p(T) = m_p c^2 \int_0^{T/m_p c^2} du/F(u).$$

$$R_M(T_M) = (Mc^2/z^2) \int_0^{T_M/Mc^2} du/F(u).$$

$$= (M/z^2 m_p) m_p c^2 \int_0^{T_M/Mc^2} du/F(u)$$

$$= (M/z^2 m_p) R_p(m_p T_M/M).$$

13.2 $T = [Mc^2/\sqrt{(1 - v^2/c^2)}] - Mc^2$

gives

$$\frac{v^2}{c^2} = 1 - \frac{1}{(1+u)^2}, \quad \text{where } u = T/Mc^2.$$

Taking a constant $\bar{L}$, the integration is straightforward.

13.3 α-particle range $\approx 20\ \mu$m; electron range ~ 1 cm, a much greater distance.

13.4 $$\frac{d}{dx}\left(\tfrac{1}{2}Mv^2\right) = Mv\frac{dv}{dx} = M\frac{dx}{dt}\frac{dv}{dx} = M\frac{dv}{dt}.$$

From equation (13.4), in the approximation $L = \bar{L}$,

$$\frac{dv}{dx} = \frac{\text{constant}}{v^3} \quad \text{and} \quad \frac{dv}{dt} = \frac{\text{constant}}{v^2}.$$

Hence

$$(\text{time to stop})/(\text{range}) = \int_0^{v_0} v^2\,dv \Big/ \int_0^{v_0} v^3\,dv = 4/3v_0.$$

For the α-particle of question (13.3), time $\approx 1.7 \times 10^{-12}$ s.

13.5 The kinetic energy of ionising particles is 0.76 MeV. From the end of § 13.1, the number of ion pairs produced is

~ 0.76 MeV/35 eV $\sim 2 \times 10^4$.

The proton will have the longest range. The proton energy is $(m_t/(m_p + m_t)) \times 0.76$ MeV $= 0.57$ MeV.

To estimate its range take $I = 24$ eV (Problem 13.3) and estimate $\bar{L}$ (~ 2.5), which gives a range ~ 0.5 cm.

13.6 From Fig. 13.4, at 50 keV the photon cross-section for lead is predominantly due to absorption and the linear attenuation coefficient $\mu \approx 93.2\ \text{cm}^{-1}$.

For the thickness x to be such that $e^{-\mu x} = 10^{-3}$, we require $x = 0.76$ mm.

Appendices

B.1(*b*) The centre-of-mass coordinate $\mathbf{R} = (m_1\mathbf{r}_1 + m_2\mathbf{r}_2)/(m_1 + m_2)$, and the relative coordinate $\mathbf{r} = \mathbf{r}_2 - \mathbf{r}_1$. The two particle wave-functions $\exp(i\mathbf{k}_1 \cdot \mathbf{r}_1)\exp(i\mathbf{k}_2 \cdot \mathbf{r}_2)$ and $\exp(iK \cdot \mathbf{R})\exp(i\mathbf{k} \cdot \mathbf{r})$ must be identical. The result follows on equating coefficients for $\mathbf{r}_1$ and $\mathbf{r}_2$. The Jacobian of the transformation is unity. If the particles are identical, only one hemisphere of the angular integration of the **k**-vector gives distinct states, since **k** and $-\mathbf{k}$ are equivalent.

C.2 The wave-functions of the state are linear combinations of spatial functions of fixed l, each of which has parity $(-1)^l$. The effect of the parity operator on the internal states $|+\tfrac{1}{2}\rangle, |-\tfrac{1}{2}\rangle$ of spin $\tfrac{1}{2}$ fermions (e.g. electrons, protons) is in fact a matter of convention, and they are taken to have positive parity. It is usually the relative parity of two states which is significant.

E.1 9×10^{16} Bq, 2×10^4 Bq m^{-2}

E.2 6 Ci.

E.4 12 μSv.

Index

(P) *denotes the reference is to a problem*